Global Issues Series

General Editor: **Jim Whitman**

This exciting new series encompasses three principal themes: the interaction of human and natural systems; cooperation and conflict; and the enactment of values. The series as a whole places an emphasis on the examination of complex systems and causal relations in political decision-making; problems of knowledge; authority, control and accountability in issues of scale; and the reconciliation of conflicting values and competing claims. Throughout the series the concentration is on an integration of existing disciplines towards the clarification of political possibility as well as impending crises.

Titles include:

Brendan Gleeson and Nicholas Low (*editors*)
GOVERNING FOR THE ENVIRONMENT
Global Problems, Ethics and Democracy

Roger Jeffery and Bhaskar Vira (*editors*)
CONFLICT AND COOPERATION IN PARTICIPATORY NATURAL RESOURCE MANAGEMENT

Ho-Won Jeong (*editor*)
GLOBAL ENVIRONMENTAL POLICIES
Institutions and Procedures

W. Andy Knight
A CHANGING UNITED NATIONS
Multilateral Evolution and the Quest for Global Governance

W. Andy Knight (*editor*)
ADAPTING THE UNITED NATIONS TO A POSTMODERN ERA
Lessons Learned

Graham S. Pearson
THE UNSCOM SAGA
Chemical and Biological Weapons Non-Proliferation

Andrew T. Price-Smith (*editor*)
PLAGUES AND POLITICS
Infectious Disease and International Policy

Michael Pugh (*editor*)
REGENERATION OF WAR-TORN SOCIETIES

Bhaskar Vira and Roger Jeffery (*editors*)
ANALYTICAL ISSUES IN PARTICIPATORY NATURAL RESOURCE MANAGEMENT

Simon M. Whitby
BIOLOGICAL WARFARE AGAINST CROPS

Global Issues Series
Series Standing Order ISBN 978-0-333-79483-8
(*outside North America only*)

You can receive future titles in this series as they are published by placing a standing order. Please contact your bookseller or, in case of difficulty, write to us at the address below with your name and address, the title of the series and the ISBN quoted above.

Customer Services Department, Macmillan Distribution Ltd, Houndmills, Basingstoke, Hampshire RG21 6XS, England

Analytical Issues in Participatory Natural Resource Management

Edited by

Bhaskar Vira
University Assistant Lecturer in Environment and Development
Department of Geography
University of Cambridge

and

Roger Jeffery
Professor of Sociology of South Asia
University of Edinburgh

Softcover reprint of the hardcover 1st edition 2001 978-0-333-79276-6

First published 2001 by
PALGRAVE
Houndmills, Basingstoke, Hampshire RG21 6XS and
175 Fifth Avenue, New York, N. Y. 10010
Companies and representatives throughout the world

PALGRAVE is the new global academic imprint of
St. Martin's Press LLC Scholarly and Reference Division and
Palgrave Publishers Ltd (formerly Macmillan Press Ltd).

ISBN 978-1-349-41942-5   ISBN 978-1-4039-0767-7 (eBook)
DOI 10.1057/9781403907677

This book is printed on paper suitable for recycling and
made from fully managed and sustained forest sources.

A catalogue record for this book is available
from the British Library.

Library of Congress Cataloging-in-Publication Data
Analytical issues in participatory natural resource management /
edited by Bhaskar Vira and Roger Jeffery.
    p. cm. — (Global issues)
    Papers presented at a workshop held at Mansfield College,
Oxford on April 6 and 7, 1998.
    Includes bibliographical references and index.

    1. Natural resources—Management—Congresses.
    2. Environmental policy—International cooperation–
–Congresses. I. Vira, Bhaskar. II. Jeffery, Roger. III. Series.
    HC85 .A53 2001
    333.7—dc21
                                        2001021198

10   9   8   7   6   5   4   3   2   1
10   09 08 07 06 05 04 03 02 01

Contents

Part III Attitudes and Responses of Intervention Agents

List of Tables, Figures, Maps, Boxes and Appendices

vii

Preface and Acknowledgements

Attempts to manage natural resources through collaboration rather than competition, through negotiation rather than by fiat, by agreement rather than conflict, have become a touchstone for many who see these efforts as the harbinger of global sustainable development. But thoroughgoing, independent assessment of the successes and failures of experimental programmes around the world, and the conditions within which such efforts might be made with some chances of success, are still in their infancy. In April 1998 over 120 academics and consultants, with representatives of NGOs and governments, came together to discuss papers that addressed these themes. It was an enjoyable three days, with considerable crossing over of the boundaries sometimes found between people from these different constituencies. Debate and discussion went on well into the night, generating ideas for new research, for changed practices within donor agencies, and for new ways to think about some of the key issues.

This book and its partner (Roger Jeffery and Bhaskar Vira, 2001, eds, *Conflict and Cooperation in Participatory Natural Resource Management*) bring together some of the papers presented at the Oxford conference. We have attempted to convey something of the excitement generated there, by making these papers more widely available. Inevitably, in the process of revising, editing, updating and preparing for publication, some of the papers have been amended considerably, in part to take account of changes since 1998 in the changing political, economic and social circumstances in the many different parts of the world considered here. In some cases (such as in Indonesia) the political changes have been so considerable that predictions based on material before 1999 offer a poor guide to opportunities in the future. None the less, we believe the chapters provide valuable insights into activities, structures and processes with great relevance for the twenty-first century.

We are grateful to many people for their work and support in making the conference possible and helping to see the project through to publication. The Economic and Social Research Council's grant funded the conference by supporting the administrative costs and allowing us to invite scholars and consultants from India, Africa and Latin America. Alistair Scott, in the office of the Global Environmental Change Initiative in Sussex, was particularly supportive of our efforts. Much of

viii

the preparatory work for the conference was done by Esther Dermott, and liaison with Mansfield College was sensitively handled by Anne Maclachlan. Keynote talks given at the conference by Melissa Leach, Iain Scoones and Ian Swingland helped to get the conference off on a sound footing. Discussants were Darrell Posey (Mansfield College), Catherine Locke (University of East Anglia), Philip Woodhouse (University of Manchester) and Helle Qwist-Hoffman (FAO). Jim Whitman encouraged us to submit a proposal for publication to Palgrave as part of the Global Issues Series, and Karen Brazier and Eleanor Birne coordinated the project for the publishers. Valuable assistance with editing the papers was provided by Lucy Welford in Cambridge and Colin Millard in Edinburgh. The Centre for South Asian Studies at the University of Edinburgh provided financial support for the editing. None of the above are responsible for the views expressed here, nor for any errors and omissions that remain.

As with all such activities, the families of the editors have suffered in various ways: we are grateful for their tolerance and aware of our debts to Shiraz and Kartik (BV) and Tricia, Laura and Kirin (RJ).

Notes on the Contributors

Margie DeWeese-Boyd received an MA and PhD in Political Science from the University of Missouri-St. Louis. She also received an MSW from the George Warren Brown School of Social Work at Washington University in St. Louis, where she currently is a PhD candidate. She has served as managing editor for *Utopian Studies*, a scholarly journal published by the Society for Utopian Studies, and has recently authored 'The State, the Market, and Civil Society: Building Sustainable Communities in Transitional Economies', with Gautam Yadama (in *Challenges of Transformation and Transition from Centrally Planned to Market Economies*, ed. Kempe Ronald Hope, Sr., UNCRD Research Report Series No. 26, 1998). DeWeese-Boyd is an Assistant Professor of Social Work at Gordon College in Wenham, Massachusetts.

Man-Kwun Chan is a sociologist, who has contributed to the development of stakeholder analysis.

Czech Conroy is a senior socio-economist at the Natural Resources Institute, University of Greenwich, and has been working on sustainable development and livelihood issues for 15 years. He was co-editor of *The Greening of Aid: Sustainable Livelihoods in Practice* (1988). As well as being involved in participatory forest management, he is currently working on participatory technology development in the livestock sector in India.

Diny van Est is at the Centre of Environmental Science at Leiden University.

Chip Fay is a senior tenure specialist working on Natural Resource Strategies and Policy with ICRAF Indonesia. From 1992 to 1995 he was a Ford Foundation programme officer responsible for developing, managing and evaluating the Foundation's activities on community management of forest lands in Indonesia. Between 1987 and 1992, he worked with the Environmental Policy Institute as Director of their Southeast Asia office. He has also worked with Survival International, where he developed and implemented programmes for Southeast Asian countries.

Hubert de Foresta works with ICRAF Indonesia.

Urs Geiser gained an MSc in Geography at the University of Zurich (1975). He then worked with development projects of the Swiss Development Cooperation (SDC) in the Yemen Arab Republic (regional planning and mapping) and in Sri Lanka (natural resources surveys). From 1983 to 1988, he was a consultant for SDC and other agencies with emphasis on South Asia (planning, monitoring and evaluation regarding projects on rural development, forestry, irrigation, livestock, etc.). He joined the Department of Geography at Zurich University in 1988 and completed his PhD in 1993 on conflicts between indigenous and exogenous concepts of resource management in Sri Lanka. Since 1993, he has been Lecturer and Research Associate at the Department. His research is on land resource management with an actor-oriented perspective, especially in Kerala, Sri Lanka and Switzerland, and he continues to be involved with development agencies (especially in Pakistan).

Ingvild Harkes is an anthropologist and currently works at the Institute for Fisheries Management and Coastal Community Development in Denmark. Previously, she has worked at ICLARM (International Center for Living Aquatic Resources Management) in the Philippines. Harkes has carried out a study on *sasi*, a traditional local fisheries management system in Maluku province, Indonesia. Currently she is working on co-management issues in West Africa where, together with local partners, hypotheses on co-management are tested and an institutional analysis of fisheries co-management arrangements are carried out.

Roger Jeffery has been Professor of Sociology of South Asia at the University of Edinburgh since 1997. His research in India since 1971 has covered health policy-making as well as village-based social demographic fieldwork in north India. He was a principal investigator for an ESRC-funded project looking at Joint Forest Management in four Indian states (1994–97). Recent publications include: *The Social Construction of Indian Forests* (ed.), 1998 and (with N. Sundar (eds)), *A New Moral Economy for India's Forests?*, 1999.

John Kabamba works in the Department of Forestry and Beekeeping for the Government of Tanzania. He is a District Beekeeping Officer for Muheza and has been seconded to the Tanga Coastal Zone Conservation and Development Programme as their forestry adviser. He has considerable experience in forestry, particularly rural communication and extension work.

Pradeep Khanna is a member of the Indian Forest Service, based on Gujarat.

Abha Mishra is a consultant on natural resource management and rural development, based in Orissa. She has been involved in research projects in India on various aspects of forest management and utilisation by local people.

Koos Neefjes has been Oxfam GB's Policy Adviser Environment & Development since early 1992. Before that he worked as technical adviser to land and water management projects in Bangladesh and Guinea-Bissau and undertook short-term advisory missions to several other countries. He was also briefly involved in postgraduate teaching in land drainage in The Netherlands and Brazil. His work with Oxfam focuses on the practical relationship between poverty and environmental change, in humanitarian emergencies and in rural and urban development. Central to that work are questions of food security and agriculture, but he also engages with urban questions on housing and public infrastructure. He has written a large number of appraisal and evaluation reports, contributed to several books and policy documents and published a number of articles on practice and policy issues. He recently published *Poor People's Environments: A Development Professional's Guide to Improving the Sustainability of Livelihoods*.

Mike Nurse is a natural resource management specialist with particular interest in participatory forest management, and is based in Canberra. He was Forestry Adviser on AusAID's Nepal-Australia Community Forestry Project between, and worked for BirdLife International in Cameroon on an integrated conservation and development project. Most recently, Nurse has been working Royal Government of Bhutan, serving as a training and extension specialist on the Third Forestry Development Project.

Gerard Persoon is at the Centre of Environmental Science at Leiden University.

Ajay Rai has worked with and researched forest management communities in Orissa and has published reports, papers and articles on the subject.

V. Santhakumar is at the Centre for Development Studies in Thiruvananthapuram.

Madhusree Sekhar is Assistant Director at the Decentralised Governance and Planning Centre at the Institute for Social and Economic Change in Bangalore. His major areas of research interest are institutions in development and their impact on local governance and environmental management. She has worked extensively in these areas and her pervious publications include: 'Analysing Urban Solid Waste in Developing Countries: A Perspective on Bangalore', 'Organisation for Sustainable Common Property Development', *Review of Development and Change,* 2 (1997); 'Panchayat Reforms in Tamil Nadu: A Historical Perspective', *Journal of Rural Development,* 17 (1998); and Working Paper No. 24, CREED Working Paper Series, IIED London (1999).

Neera M. Singh has worked and researched forest management communities in Orissa and has published reports, papers and articles on the subject.

Nandini Sundar is Reader in Sociology at the Institute of Economic Growth, Delhi. Her previous publications include: *Subalterns and Sovereigns: An Anthropological History of Bastar,* 1997; and (with Roger Jeffery (eds)) *A New Moral Economy for India's Forests: Discourses of Community and Participation,* 1999. She was a member of the Edinburgh University/ICFRE team studying JFM in India.

Bhaskar Vira is university assistant lecturer in Environment and Development at the Department of Geography, and a Fellow of Fitzwilliam College, at the University of Cambridge. His research examines the political economy of environmental and natural resource management in the developing world, with a special focus on forestry issues in India. He is particularly interested in issues of governance and the interaction of multiple stakeholders (especially government agencies, NGOs and local groups) in the context of resource use and management. Previous publications include: 'Institutional Pluralism in Forestry: Considerations of Analytical and Operational Tools', *Unasylva* 49 (1998); and 'Implementing Joint Forest Management in the Field: Towards an Understanding of the Community-Bureaucracy Interface', in R. Jeffery and N. Sundar (eds.) *A New Moral Economy for India's Forests? Discourses of Community and Participation,* 1999.

Gautam Yadama is an associate professor, and coordinator of the Social and Economic Development Concentration at the George Warren Brown School of Social Work at Washington University in St. Louis. His research has addressed issues related to poverty, the role of non-governmental organisations in sustainable development, and self-governance of common pool resources. His particular interests are in understanding how indigenous communities work with the state to manage forests that are under increasing pressure from state and non-state entities. His current research focuses on understanding exogenous pressures on micro-institutional mechanisms for managing community forests in marginal communities in India, Bhutan, and Turkey. He is co-author with Arun Agrawal of an article in *Development and Change* titled 'How do Local Institutions Mediate Market and Population Pressures on Resources? Forest Panchayats in Kumaon, India'.

1
Introduction: Analytical Issues in Participatory Natural Resource Management

Bhaskar Vira and Roger Jeffery

1. Participatory natural resource management: theory and practice

The participatory perspective has informed a number of policy initiatives in natural resource management in the developing world, particularly over the past two decades. There is a growing recognition that top-down projects have limited potential for transforming existing patterns of social interaction and resource use, because they do not relate adequately to local priorities. The received wisdom suggests that participatory natural resource management projects work because traditional knowledge of the resource and existing social structures can be utilised to develop more effective strategies for resource use. As a result, participation is accepted as an integral part of the 'new paradigm' of development which is being promoted by multilateral and bilateral donors in their interactions with governments in the developing world. Many developing country governments are not averse to adopting these approaches, although there is considerable variation in the extent of their commitment to policies that may be seen to challenge their control over resources and territories previously in the public domain. Donors are frequently in a good position to promote new ideas, helped by the leverage available by making project sanctions conditional upon such changes. However, there is substantial scope for the dilution of what may be seen as a potentially radical agenda through the everyday processes of implementation that characterise state action in the developing world. Perhaps these implementation 'failures' can be seen as the new 'weapons of the weak' employed by developing country governments, as well as other field-level agents, in response to what could be seen as a current, possibly fleeting, preoccupation with participation.

The widespread acceptance of the participatory approach by development practitioners – governments, non-governmental organisations (NGOs), as well as donors – suggests, however, that it may be inappropriate to dismiss it as yet another passing fad. The large volume of academic literature on participation that has emerged in recent years certainly suggests that the intellectual community is engaging seriously with this concept. If the belief that participation represents an abiding design principle for the present and next generation of development interventions is widely shared, perhaps there is some merit in subjecting the concept to close scrutiny. The chapters in this book, all originally presented at a conference held in Oxford in April 1998, critically review the theory and practice of participatory natural resource management in the developing world. They examine the participatory process from a variety of perspectives, drawing general conclusions from detailed knowledge of specific empirical situations. The book is organised into four parts, each of which deals with broad themes that emerged during the conference and the subsequent discussion. Each of the chapters relates primarily to the specific themes of its part, but the discussion is relevant to other themes as well.

The first major theme relates to conceptual and analytical issues in the participatory context. Geiser, and van Est and Persoon, engage with the way in which concepts such as community, resources and time are used by social agents to inform their claims over the participatory process. Participation must be understood as a dynamic and contested term, which needs to be contextualised in space and time. Part II recognises that participation is an evolving process and that collaboration often requires purposive action. The chapters by Yadama and DeWeese-Boyd, Nurse, and Sekhar demonstrate the manner in which participatory organisations and institutions need to be constructed, but can also be destroyed in particular circumstances. The third major theme deals with the way in which agencies and organisations, within both the governmental and non-governmental sectors, implement and interpret participatory initiatives. The process of managing and evaluating participatory projects is discussed by Neefjes and Harkes, while Jeffery *et al.* demonstrate the manner in which attitudes of implementing agencies and their staff vary with respect to their own personal circumstances, as well as the nature of the planned intervention. The final part discusses the dynamics of participation, especially the emergence of conflict and its management, as well as the processes by which specific policies evolve and persist. Conroy *et al.* focus on the emergence of conflict at different levels, and the way in which this can be managed by specific policy and institutional structures. Fay and de Foresta, and Santhakumar pay particular attention to the policy process, and the manner in which social actors are able to influence the direction of participatory initiatives, as well as their persistence over time.

2. Analytical issues

It is now well recognised that participation can take a variety of forms, ranging from relatively superficial information sharing to full empowerment (Midgley 1986; Paul 1987; Pimbert and Pretty 1997; UNDP 1997). These typologies are useful in providing an analytical and descriptive template for organising discussions about participation. They also reflect an implicit hierarchy of participatory attitudes and outcomes, with those at the 'lower' end of such scales being seen as less worthy than the more enlightened examples at the 'top' end. Jeffery *et al.*, however, point out that these attitudes need not be mutually exclusive. Within any single organisation, different individuals may have different attitudes and responses to participatory initiatives; equally, individuals themselves may adopt different positions simultaneously, perhaps with reference to different aspects of the implementation process. Geiser suggests that 'participation can be understood as the purposive interaction of social actors with other social actors with a view to achieving specific outcomes'. In this account, participation is negotiated and contested, and the critical issue is who participates with whom (against whom) for what purposes?

Participatory approaches bring groups and individuals together to work towards the management of a set of resources. Recent literature suggests that it may be useful to think of such social actors as stakeholders, defined as 'all those who affect, and/or are affected by the policies, decisions, and actions of the system: they can be individuals, communities, social groups or institutions of any size, aggregation or level in society' (Grimble *et al.* 1995; Grimble and Wellard 1997). Conroy *et al.* (chapter 10) distinguish further between primary and secondary stakeholders, where the former are those who have more direct interaction with the resource. The process of negotiating participatory strategies may sometimes involve the dilution of the particular interests of specific social actors, in order to achieve an acceptable compromise (Vira *et al.* 1998). However, the chapters by Conroy *et al.* and Neefjes (chapter 7) both point out that since the distribution of power between stakeholders is unlikely to be equal, this can lead to the systematic exclusion or neglect of the interests of weaker sections. Furthermore,

Geiser (chapter 2) and Neefjes warn against too narrow an interpretation of stakeholders, suggesting that development interventionists and researchers need to recognise themselves as among those whose actions impact on resource management outcomes.

It seems increasingly apparent that participatory strategies are adapted and moulded by stakeholders to reflect their specific needs and interests. Geiser and Neefjes, for instance, draw on the work of Long and Long (1992), and highlight the role of social agency, suggesting that it is far too simplistic to conceive of groups merely as passive beneficiaries of participatory interventions. Santhakumar (chapter 12) and Sekhar (chapter 5) both use frameworks from the new institutional economics which suggest that individual responses are shaped by incentives that emerge from the physical, technical, economic and institutional circumstances in which they are embedded. Jeffery *et al.* (chapter 9) document this in their study of bureaucratic attitudes towards participatory forestry in India, finding that individual actors do indeed have considerable autonomy in their reactions to such initiatives. The important analytical insight is that participatory outcomes may be influenced to a significant extent by the actions of specific agents. In practical terms, this suggests that project planners must take account of the role of social (and individual) agency in shaping the participatory process.

The literature on participation has evolved considerably in recent times, and has recognised that the notion of 'community' is a complex one. It is certainly widely acknowledged that communities are not simple homogeneous groups that work harmoniously to promote group objectives. In some senses, this literature has come full circle: from early pessimism about community action as exemplified in the work of Hardin (1968), to the relatively uncritical view which informed community-based conservation initiatives through the 1990s (Western and Wright, 1994), to a more subtle and nuanced current understanding of communities as complex and dynamic, and characterised by internal differences and processes (Leach *et al.* 1997a; Agrawal and Gibson 1999). The chapters in this volume adopt a more critical analysis of community. Geiser (chapter 2), for instance, suggests that the community as a target beneficiary of participatory approaches may change over time, as different interventions adopt various strategies to promote particular ends. Communities may also reconstitute themselves, spontaneously or in response to external opportunities. Van Est and Persoon (chapter 3), for instance, describe the process of mobilisation of the Bugkalot community, who did not previously define themselves as a corporate social group, in order to benefit from specific governmental action in the Philippines. What these examples suggest is that researchers have to be very careful about the context in which they use the idea of community. Acknowledging these critiques, Conroy *et al.* adopt a very specific definition of community, based on the work of Ostrom (1992), in which they restrict the term to imply groups that are characterised by shared beliefs, stability of membership and complex, multilayered, long-term interaction.

The chapters in this volume subject a number of other analytical generalisations to critical scrutiny. Geiser suggests that there may not be commonly shared beliefs about what constitutes the resource that is to be the subject of participatory efforts. Extending the idea of differentiated interests, he suggests that 'different actors may perceive ... specific components as resources suitable, or required, for their economic activities. ... Analysis then has to ask which component ... is considered by whom (at which point of time) as a resource, and for what purpose.' Van Est and Persoon also comment on the need to focus on the nature of the resource, since '[d]ifferent resources require different kinds of management regimes'. Extending models of management that are relatively effective in the context of a particular set of resources may be inappropriate in situations where the physical characteristics are substantively different. Institutional analysts are conscious of such differences (Thompson and Freudenberger 1997; Ostrom 1999), but single resource case studies tend not to highlight them to the same extent as they do variations in community characteristics, or in rules and regimes for resource management. Geiser's work goes beyond such literature, by suggesting that resource characteristics are not just physical, but are socially constructed. He documents the changing and contested perceptions of the forest resource in Kerala, India, where it has been viewed at different times as a specific land use which competed with agriculture, as a source of industrial raw material, as a source of non-timber forest produce for tribals and settlers, as well as an ecologically significant system. Social actors use these particular constructions of the forest to reinforce their claims for privileged access to, and control over, the fate of this resource.

Van Est and Persoon suggest that specific stakeholders may use constructions of time as tools to reinforce their claims over resources. They suggest that the future is central to current discourses about sustainability, used either by those who predict impending ecological catastrophe to justify interventions, or by those who construct images of a desirable configuration of social and ecological systems in order to justify current policy actions. Local resource users employ a different terminology, justifying their claims because of previous ownership or stewardship of the resource, and also setting up a conflict between the essential survival needs of the present generation of the poor against the luxury of environmental concern as manifested in post-materialist discourses as well as those that refer to the interests of future generations (Guha and Martinez-Alier 1997). The development bureaucracy and project planners have a very specific time perspective, reflecting both the periodicity of project funding, planning, review and termination, as well as the political and electoral cycles. Analytically, what is interesting is the manner in which these different perspectives on time come together and interact in the context of specific participatory initiatives. Van Est and Persoon suggest that the temporal dimension needs to be more explicit in the study of such interventions, and illustrate the utility of such a framework using two examples from the Philippines and Cameroon.

3. Constructing participation

A second set of issues that emerge from this volume relate to the manner in which participatory strategies evolve and are constructed. Sekhar suggests that interventions may be justified by the need to alter a structure of incentives that is leading to the over-exploitation of resources. While it is clear from the previous discussion that perceptions of resource exploitation and degradation may not be universally shared, they are nevertheless frequently invoked as a reason for purposive action, especially by external agents. What such strategies seek to construct is a social, economic and political context within which social agents can collaborate with each other in pursuit of specific objectives. There is a growing awareness that single agencies may be unable to manage resources in a manner that is sensitive to the needs of multiple stakeholders (Anderson *et al.* 1998). The recognition of the inadequacy of previous, frequently state-dominated, strategies of resource management has been coupled with a large literature that has highlighted the potential for partnerships that involve other stakeholders, especially local resource users (Baland and Platteau 1996).

While some evidence exists to suggest that local resource users are able to respond spontaneously to changing social and physical circumstances and to evolve effective resource management strategies, collaborative arrangements need to be consciously created by purposive action. Participation can be seen as an example of co-production (Ostrom 1996) or shared management, where these are defined as regimes 'which produce goods or services by utilising inputs from (at least two) individuals or legal entities who are not part of the same organisation, and are not under the control of a single principal' (Vira *et al.* 1998). If these social agents have not worked together previously, or even more importantly if they have a history of conflict, participatory strategies have to be actively promoted. Yadama and DeWeese-Boyd (chapter 6), for instance, document the manner in which the Indian forest department and tribal communities have contested issues of resource use and access at least since colonial times. Despite this history, India's Joint Forest Management (JFM) programme recognises the potential for these groups to collaborate in the care and management of forest resources, and consciously attempts to create an institutional context within which this is possible.

Collaborative approaches need to build on existing social arrangements, and to create new ones if those that exist are inappropriate. Fay and de Foresta (chapter 11) refer to the abundance of ecologically sound agroforestry systems that are found in Indonesia, and the manner in which the rights of these community-based systems have recently been recognised within the State Forest Zone. They see this as 'a victory for those in Indonesia who were encouraging the Department of Forestry to use what is already being done by local farmers as the basis or starting point for community-oriented forest management'. Where indigenous systems of resource use and control are already in existence, these provide an obvious local organisational structure with which other stakeholders can collaborate. From their case study of Tanzania, Nurse and Kabamba (chapter 4) conclude that externally sponsored institutions which are created in situations where such local structures do not exist may nevertheless be reasonably robust and provide a sound, as well as equitable, basis for local decision-making. They highlight the need for continued external support to ensure the stability of such interventions, and the need for policies to evolve to create an enabling structure for the wider replication of such strategies. However, they stress that the 'organisational and institutional structure for forest conservation under this scenario will be uniquely Tanzanian (and unique to each site in Tanzania) and needs to be responsive to a rapidly changing political, economic and cultural environment'.

In her study of Joint Forest Management (JFM) in Orissa, in India, Sekhar contrasts three different types of local rural organisations; one is promoted by the government, the other has emerged due to action by a local NGO, while the third is a 'traditional' management regime constituted without external intervention. Sekhar suggests that each of these regimes had specific advantages in the context of organising local participation: the government-sponsored organisation was seen as legitimate and ensured accountability, but did not leave much scope for local autonomy and initiative; the NGO was an effective mobiliser, but seemed less able to secure the long-term sustainability of the chosen strategies; the traditional regime had substantial community support, but there was a risk that the consensus-based decision-making system could break down. Sekhar concludes that an appropriate strategy would be to combine these different organisational perspectives into an integrated approach, which was able to draw on the synergies between local and externally sponsored organisations.

Yadama and DeWeese-Boyd focus on the supply of social capital – shared norms, trust and information – and the manner in which this impacts upon participatory initiatives. Since social capital is rarely created consciously, they suggest that it probably emerges as the by-product of other social activities, and point to the existence of civic networks, or the process of negotiating resource use conflicts, as potential sources of social capital. They use the example of an NGO working in tribal villages in Andhra Pradesh, India, which established its local credibility by mediating on behalf of tribals in land conflicts. This created a sense of trust and reciprocity – social capital in their terms – that the NGO could leverage in the context of other programmes, including JFM. JFM requires the Forest Department and tribals to learn to trust each other to work towards their mutual benefit; in the case under review, the presence of a credible NGO helped overcome suspicions on both sides, especially on the part of the tribals. However, collaborative strategies of this sort can also prove to be fragile, since they depend heavily on the presence of external facilitators. In their case study, Yadama and DeWeese-Boyd show that the rise of armed militancy in these areas has forced the withdrawal of the NGO, undermining JFM arrangements between the Forest Department and tribal communities.

The example used by Yadama and DeWeese-Boyd demonstrates the important role that NGOs can play in facilitating participatory strategies, particularly in situations where the collaborating parties do not have full confidence in each other. Jeffery *et al.* display the manner in which the role of NGOs is formally set out in the participatory forest policy statements of the government of India, but also point to the ambivalent attitude of some Forest Department staff to such involvement. Fay and de Foresta discuss the manner in which NGO and academic networks are collaborating in Indonesia to secure the natural resource rights of forest-based communities by altering the policy and institutional framework. Conroy *et al.* draw on empirical material from Orissa in India which supports the need for NGO involvement at levels other than that of local project implementation. They suggest that lower-level organisations need to be supported by multi-stakeholder decision-making fora at all levels – local, district, provincial and national. Community organisations need to be part of a structure of 'nested' enterprises (Ostrom 1999), such as apex organisations and federations, in order to coordinate their activities, assist with conflict management and provide an interface for dealings with the Forest Department. Conroy *et al.* especially emphasise the role that NGOs can play in situations of conflict, since they are seen as neutral third parties who are 'able to break situations of deadlock and create an environment for the conflicting parties to come to negotiations'.

4. The attitudes and responses of intervention agents

The processes by which participatory projects are planned, implemented and evaluated clearly affect outcomes, but also have significant dynamic effects on implementing agencies and their capacity to learn from experience. Neefjes (chapter 7) draws on the experience of a large development agency, Oxfam GB, to illustrate the manner in which participatory approaches impact on project management. The organisation has adopted a sustainable livelihoods framework as an instrument for developing strategic plans, setting objectives and improving the understanding of linkages between environmental change and poverty alleviation at the micro-level. The framework does not *require* the participation of stakeholders in the analysis, but participatory approaches are increasingly being stressed in current practice of impact assessment. In the case of Oxfam, this has taken a hybrid form, known as Participatory Environmental Assessment, in which the sustainable livelihoods framework is used to analyse the context, while participatory tools are combined with environmental (impact) checklists for monitoring and evaluation. Although there is little hard evidence which would allow us to assess the impact of such methodologies, Neefjes finds that project reviews and anecdotal feedback suggest a 'limited and always positive' impact on natural resources in projects where the hybrid approach was introduced.

Harkes (chapter 8) observes that frequently there are differences in the perceptions of project staff and beneficiaries about the extent of project success. She suggests that the perspectives of project participants and those of external agents do not coincide, and this has important consequences for the analysis of project outcomes. Participatory methods have a potentially transformative role in terms of the personal capabilities and attitudes of the participants, who 'need to acquire communication skills, the ability to formulate project goals, to plan and delegate tasks, solve problems and other abilities necessary to be a partner in executing a community-based management project, a process usually referred to as capacity building'. These skills are usually developed in the pre-implementation phase of a project, and are vital for beneficiaries to understand project interventions and to implement them. However, the nature of the project cycle often means that material interventions are carried out while beneficiaries are still in the process of developing these skills, and this can result in a negative evaluation. Furthermore, most evaluations tend not to include the personal achievements of participants as indicators, focusing instead on quantitative and material indicators of project success. Harkes concludes that 'the reason why projects are often evaluated as being unsuccessful is because neither the timing of the evaluation, nor the criteria used to measure success are appropriate'.

Harkes suggests that there should be two stages at which project impacts are evaluated. While social preparation is underway, evaluations should concentrate on the internal or 'emic' level, which consists of 'elements, aspects, and interpretations of the belief system as perceived or conceived by the members of the culture or society under consideration'. Only once the project has matured and project interventions have been carried out is evaluation according to external or 'etic' criteria appropriate. The early emic evaluation allows projects to adjust goals and strategies in response to local perceptions, and thereby prevents overall project failure. Furthermore, if beneficiaries develop project goals as equal partners with project implementers and formulate objectives collaboratively, the internal and external expectations from projects should coincide, and one can expect that the definition of success for the implementers and the participants would be similar. The implications for project proponents are significant, since Harkes' analysis suggests the need for 'a longer preparatory process, a redefinition of project goals, and possibly a longer implementation period'. This challenges not only existing structures of project planning and funding, but also the extent to which external agents can exercise their authority over projects.

The chapters by Neefjes and Harkes both have profound implications for our understanding of the nature of participatory strategies.

If participation is to be deeply embedded in project interventions, it can be seen to extend beyond implementation to processes of consultative planning and project design. As a result of such consultation, goals and objectives of interventions may eventually reflect priorities of stakeholders other than the implementing agency, and this may have to be accepted as a necessary dilution of these initial objectives. However, as Neefjes reminds us, funding and control issues mean that the project implementers continue to enjoy a relatively privileged and powerful position relative to other stakeholders, and this is likely to limit the extent to which the participatory process can be viewed as egalitarian. Similarly, Conroy *et al.* warn that 'the stronger stakeholders can be expected to dominate decision-making; and, when conflicts arise, to promote their interests above others'. The manner in which implementing agencies use this power would vary, with some organisations actively seeking to learn through external consultation and internal self-reflection, while others remain firmly committed to their original objectives and may be relatively inflexible about their response to the perspectives of other stakeholders. In reality, many organisations, both government agencies and NGOs, probably have attitudes and structures that lie in between these extremes and this would be reflected in the extent to which beneficiaries are consulted in the processes of project planning, monitoring and evaluation.

This diversity of organisational responses to participatory initiatives is documented by Jeffery *et al.* (chapter 9), who study the discourses of participation in the forest departments of four Indian states. As has been suggested previously, the history of the relationships between Indian Forest Departments and local resource users has been one of conflict, and some critics see this as inimical to the adoption of participatory approaches under JFM. On the other hand, there are numerous examples of field contexts within which forest departments have been able to forge collaborative relationships with local users (Poffenberger and McGean 1996). To study the Indian forestry administration as an undifferentiated monolith would hide these variations, both between different organisations as well as at different levels within any single structure. Jeffery *et al.* find ambivalence and dissatisfaction with service conditions among their respondents, but suggest that some of these views are not dissimilar to those of other public sector employees. Apart from these generic concerns, they find that 'the main areas of desired change ... were towards forms of working which would allow for better relationships between FD staff and forest dwellers', thus clearly seeking to reverse the previous history of conflict. As far as JFM is concerned, they report almost no opposition to the participatory rhetoric in public, although the reasons for this vary between staff at different levels of the organisations. JFM emerged from a process of policy reform that was concentrated at the most senior levels of the central government. Problems with this process are recognised by some senior staff within forest departments, who suggest that 'the "real" policy makers may not have made sufficient effort to ensure that middle management and FD field staff have a sense of "ownership" of the new policies'. Specific criticisms of JFM relate to the potential for conflict at a number of levels as a result of the changes that are being introduced under JFM, to the role of NGOs within the JFM process, and to the ability of JFM to deliver its promised benefits.

In terms of actual practice, despite the general proclamations of support for JFM, Jeffery *et al.* find that the everyday work practices of field-level and middle management staff have not changed substantially. They suggest that this is because few staff are willing to be innovative and alter established patterns of work, because of a lack of structural change within departments, and because there are few reflections of participatory practices in other departments or in the rest of Indian society. This is mirrored in the findings of Fay and de Foresta (chapter 11) on Indonesia, who report that the 'tendency has been that Department staffs...revert back to familiar, prescriptive approaches that run counter to the objectives stated in the community forestry policy framework'. In his chapter, Neefjes reports a similar reluctance among front-line Oxfam staff to adopt and use the sustainable livelihoods framework, and he suggests that this may be because the framework 'is too abstract for these members of staff'. These examples suggest that there is considerable inertia in the manner in which implementing agencies work. It is far from clear that the adoption of new participatory strategies at senior organisational levels will readily translate into new working practices at the field level.

5. Dynamics of participatory processes

The chapters in this volume all recognise that participatory strategies are embedded in a wider network of social, political and economic processes. The adoption of participation involves changes in established practices, and this may not be possible without supporting changes at other levels within the system. There is a suggestion that implementing participation is as much about the political economy of institutional change as it is about designing locally appropriate interventions.

Conroy *et al.* (chapter 10) explicitly examine alternative strategies in terms of their political feasibility, recognising that 'power relations between communities cannot be changed easily'. Their chapter identifies the different types of conflict that may emerge after the introduction of participatory strategies. While some of these would be at the local level between resource users or over territories, others may reflect a lack of compatibility between local objectives and those of higher level stakeholders, or emerge because of contradictions at the policy level. Although conflict may emerge because of new participatory strategies, there are many coping mechanisms. At the local level, Nurse and Kabamba (chapter 4) show how the problem of overlapping jurisdiction in two Tanzanian villages was overcome by agreeing a joint management structure. In the case of conflicting objectives between local and higher level stakeholders, Conroy *et al.* highlight the need for mutual accountability, especially of state agencies that are responsible for implementing participatory initiatives. They also point to the use of corporatist decision-making bodies at every level, as well as the role of community apex bodies that can increase the bargaining power of local users in their interactions with other stakeholders. If conflicts emerge in spite of such measures, Conroy *et al.* suggest that existing and new mediating mechanisms (including NGOs) can be used to manage such situations. They also argue that intervention agents need to be conscious of the potential for conflict and to incorporate capacity-building measures for conflict management into project and programme design.

There is a delicate relationship between state agencies and local resource users in the participatory context. As has been suggested previously, the history of this interaction has not been free from conflict. At the same time, participation envisages a qualitative change in this relationship. Conroy *et al.* suggest that one way in which this can be supported is by changes in formal legal or administrative frameworks. Fay and de Foresta (chapter 11) document the manner in which such changes have emerged recently in Indonesia to provide a formal basis for traditional agroforestry practices in the state forest zone. In their detailed study of this process, they point to the important role played by coalitions of NGOs and academics in drafting and revising these new policy initiatives. They distinguish, in particular, between the long-term objectives of coalitions that seek to reverse the classification of community lands as state forests, and their short-term goals to secure limited use and management rights on state forests. Importantly, they argue that the 'process by which forest policy in Indonesia is changing is closely linked to the larger process of national political transformation'. To the extent that this wider transformation is still (at the time of writing) an ongoing and volatile process, their chapter must be read as an accurate assessment of current prospects, but one which may well be challenged by deeper and more fundamental changes in the structures of Indonesian politics.

Fay and de Foresta document the complex nature of negotiations that are required in order to change formal institutional structures. Santhakumar (chapter 12), in contrast, uses case studies from Kerala, India to speculate on the factors that lead to the persistence of institutions, even if these are seen to be ineffective in a local context. His chapter studies the way in which three initiatives, two state-sponsored (group farming and irrigation) and one NGO-led (local-level participatory planning) have been less than successful in meeting their objectives, but have not been revised to take account of these observed failures. Santhakumar suggests that some of the reasons for the persistence of ineffective institutions may be found in the process by which agencies monitor and evaluate their own performance, including the role of appropriate feedback on agency actions from other stakeholders. Failures may be seen to arise because of poor implementation and not be 'directly attributed to the proposed institutional form', because agencies have an ideological commitment to specific strategies. Furthermore, he suggests that some actors within agencies may have incentives not to revise ongoing strategies, and may be in a position to ensure that current practices continue despite evidence about their lack of effectiveness. The important lesson that Santhakumar draws from his study is that external interventions, especially those such as participation, which have a strong ideological basis, should not prejudge their effectiveness in a particular local context. This requires fundamental changes in the way in which implementing agencies operate, but Santhakumar is pessimistic about the prospects for this. He argues that 'the political economy of participation, in which external funding agencies, national and state governments and non-governmental organisations attempt to institute organisational changes in supposedly traditional and backward societies, provides sufficient incentives for the continued implementation of programmes that are based on a poor understanding of what is institutionally appropriate in a local context'.

6. Conclusion

What general conclusions can we draw from the present attempt to synthesise some of the experience and learning of the last two decades of participatory projects? The volume spans a broad range of issues that emerge from the participatory discourse, from understanding and interpreting the concept itself, to the process by which partnerships are constructed and managed, as well as the wider institutional and policy structures that facilitate new strategies. What the chapters propose is a more subtle understanding of difference when analysing the way in which diverse social actors interact in the context of participatory resource management. Participation itself is constructed and interpreted differently by different agents, and it may be more appropriately viewed as a process of selective engagement between some actors to further a particular set of interests at a specific point in time. Participatory processes are affected by the actions of actors at every level, and it is necessary to recognise analysts and development practitioners as agents in this process. Communities are differentiated along many dimensions, and their responses to participatory initiatives reflect this diversity. Actors also call upon different constructions of seemingly homogeneous categories such as resources and time to reinforce their specific claims over benefits from management and, as a corollary, to exclude those of others.

There is a sense in which one can question the need to subject the participatory process to such a rigorous examination. Perhaps it is more appropriate to recognise that participation is a flexible and adaptable concept, and one that can take many different shapes to reflect local circumstances and priorities. There may, then, arguably be less merit in the academic discourse about participation (of which the present volume is one example) than in the practice of participation in action. However, we would disagree with such a passive view of the academic process. As was apparent during the course of the conference that generated these studies, there is an important need for constructive engagement between the worlds of academia and actual practice in the participatory context. This is partly reflected in the backgrounds of the contributors to this volume, spanning as they do the academic, NGO, national government and donor sectors.

Perspectives and attitudes towards participation have evolved considerably in recent years. It is now no longer necessary to make a case for the inclusion of participatory principles in project planning and management, and this can partly be seen as the result of sustained efforts on the parts of academics as well as practitioners to promote the mainstreaming of this agenda. Perhaps the greatest strength of the participatory approach is its seemingly inclusive agenda and its apparent ability to accommodate a range of different perspectives. This may also be seen by some as a significant source of weakness if existing patterns of power and structural relationships reproduce themselves, despite adopting the participatory rhetoric. The chapters in this volume provide reason for caution as well as optimism while studying participatory processes. Most importantly, perhaps, they suggest the continued need for dialogue between all those who are involved in studying and implementing these evolving social systems.

Part I
Analytical Issues

2

To 'Participate' with Whom, for What (and against Whom): Forest Fringe Management along the Western Ghats in Southern Kerala

Urs Geiser

1. Introduction

What is the meaning of 'participation' within the popular notion of 'participatory natural resource management'? An interesting entry point into this debate is Oakley and Marsden's (1984) review of discourses and practices, in which they suggest four broad understandings: (1) informing and mobilising target groups; (2) community development; (3) organising target groups; and (4) empowering beneficiaries. They clearly favoured the last two, and criticised the first two for the underlying assumption of 'so-called "communities" which operate in the "national" interest and which assume that everyone is, or should be, pulling in the same direction'. Instead, they suggest that it is necessary 'to identify particular groups with conflicting interests... [showing that] different interest groups struggle for control of available assets and resources' (1984: 9). Underprivileged groups were thus to be organised and empowered – through development projects – by outside organisations such as NGOs or specialised government agencies.

While development practice embarked upon the ventures of organisation and empowerment, some academic analysis started to question this understanding of participation, and especially the relationship between those initiating participation and empowerment, and those to be empowered. A crucial extension of the analysis was later introduced by Long and Long (1992), who wrote that 'issues of policy implementation should not be restricted to the study of... intervention by governments, development agencies and private institutions, since local groups actively formulate and pursue their own "programmes of development", which may clash with the interests of central authorities'. These groups of actors are not 'passive recipients of intervention, but *active participants* who process information and strategize in their dealings with various local actors as well as with *outside institutions and personnel'* (Long and Long 1992: 21, emphasis added; for early research along such lines, see Bierschenk 1988; Olivier de Sardan 1988).

To conceptualise 'participation' within such a broader understanding of development processes, Long and Long based their argument on Giddens' (1984) structuration theory, especially using the notion of *social agency* which attributes 'to the individual actor the capacity to process social experience and to devise ways of coping with life. ... Within the limits of information, uncertainty and the other constraints (e.g. physical, normative or politico-economic) that exist, social actors are "knowledgeable" and "capable". They attempt to solve problems, learn how to intervene in the flow of social events around them, and monitor continuously their own actions, observing how others react to their behaviour and taking note of the various contingent circumstances' (Long and Long 1992: 23, summarising Giddens 1984: 1–16). Long and Long (1992), Blaikie (1997) and Peluso (1992) all forcefully show that for a critical analysis of 'participatory natural resource management', some of the underlying notions need to be studied carefully.

1.1 People

Many development interventionists still practise participation using aggregated social categories such as 'communities' or 'small farmers'. Following Oakley and Marsden, Giddens and others, such categories are too normative and therefore need to be disaggregated – not along other preconceived ideas, but based on *actual observation of social actors' practices*. Within such an approach, development interventionists themselves (and researchers?) are studied as social actors, among others, and not left out of the analysis (an aspect that often escapes attention when the notion of stakeholders is used).

1.2 Participation

The notion of participation itself is especially challenged by the concept of social agency. Social actors, by definition, are socially oriented and interact with others with a view to 'making a difference to a pre-existing state of affairs or course of events' (Giddens 1984: 14). *Participation is understood in this chapter as the purposive interaction of specific social actors with other social actors in view of achieving specific outcomes.*

In interactive contexts, expected outputs and ways of achieving them are negotiated, though some actors may be more powerful in pressing their interests, and may use 'participation' for strategic purposes only. Important questions for analysis include: who joins hands with whom – against whom – and for which objectives? Which 'strategies' are followed by whom to make claims on specific resources which dominate the claims of others?

1.3 Natural resources

This notion needs differentiation too. From the natural stock available in a specific locality, different actors may perceive (at a given point of time) specific components as resources suitable, or required, for their economic activities. Forests, for example, can thus be perceived as a *bundle of resources* (Peluso 1992). Analysis then has to ask which component from the natural stock is considered by whom (at which point of time) as a resource, and for what purpose.

This chapter is an attempt to use such a differentiated conceptual approach to discuss past episodes of participatory natural resource management in the context of forests along the Western Ghats in the Indian State of Kerala.[1] The analysis focuses on two questions: (1) What was the nature of people's participation in natural resource management? and (2) How did different actors perceive, value and manage natural resources in view of their interests? (sections 2 and 3). Insights gained from this analysis (summarised in section 4) are then used for a socio-politically and historically informed discussion of present efforts towards people's participation in Kerala (section 5). Section 6 concludes with a discussion of alternative future scenarios for participatory natural resource management in Kerala.

2. Context of the study

The Indian state of Kerala, one of the most densely populated areas in the world, stretches between the coast of south-west India and the mountain chain of the Western Ghats. Major agro-ecological zones are the lowlands with paddy and mixed cultivation around homesteads; the midlands with the foothills of the Ghats (paddy; on higher grounds rubber, homesteads, patches of forests); and the highlands consisting of the western slopes of the Ghats, transected by many valleys (rubber, remaining forests, some tea plantations) The present study is located in the highland region). The highest areas of the Ghats are the high ranges (tea, cardamom, forests).

During the last decades, many people converted forests along the Western Ghats into agricultural land, and in order to achieve this, 'participated' with others in many ways. One such settler is P.K., and his story is narrated in the Appendix.

3. Episodes of participation

P.K. and his family belong to poorer sections of Kerala's society – but they were, and are, actively involved in using the natural resources available to them in their world up the Chittar side valley. Over time, they developed their 'livelihood package' to ensure and improve living standards. In order to achieve this, they planted cash crops and gained access to specific opportunities (assistant postman; licence to collect non-timber forest products – NTFPs; new land). They participated with others in many forms to ensure access to such opportunities, or to protect existing assets (in which they sometimes failed). At one time, they actively participated in a government scheme providing land to individuals; another time they passively participated – called (disrespectfully) freeriding by some economists – in a scheme announced by a political party, but were considered as non-participants in the Forest Department's efforts to conserve forests and evicted from parts of their land.

In this section, such episodes of participation over the years are discussed by locating the story of P.K. within wider social processes in the case study area of Chittar as well as Kerala in general. This is to understand the political and historical dynamics of 'participation' – as an input into the discussion of present efforts.

3.1   Opening up of forests

Until the Second World War, the valleys and foothills of Chittar were used by only a few tribals practising shifting cultivation and collecting NTFPs; some timber was extracted by concessionists of the state, and huge rubber estates advanced to the entrance of the valleys.

In the mid-1940s, the region faced an acute food shortage, and the authorities gave permission to clear forests in the valley basins of the Western Ghats for food production. Some localities in Chittar were included in this programme. Land was not allocated to individuals directly, but through 'non-governmental organisations'. In one valley, 200 acres were given to a Christian organisation; the neighbouring valley went to the Sree Narayana Dharma Paripalana Yogam, an organisation of the Ezhava caste. Another valley bottom went to members of the Nair Service Society (representing the Nairs), another to members of the Kurava caste, and one small valley was given to ex-servicemen of the Indian Army. The permit was temporary, allowing the clearance of undergrowth only, but not the cutting of mature trees. After some years, the settlers were expected to plant forest trees and return the land to the state. But this did not happen.

The perception of natural resources seems to have been the same by the state and settlers: *land* to expand agricultural production. The arguments justifying forest clearings were also the same: the need to provide food, work and income to a growing population. Having the same resource component in mind, and the same intentions, people (through their NGOs) fully participated in the government scheme of opening up forests. The losers were the tribals (though there were few in Chittar), who had to accept a reduction of their economic perimeter as they were interested in another component of the resource bundle, NTFPs.

3.2 Settlers establish themselves

The 1950s was a period of great transition. The state of Kerala was in its formative stage, population growth was high and there were more government schemes to provide (forested) land to settlers. Again, land was not given directly to individuals, but now to leaders of groups, who then distributed the land, as in the case of P.K.'s grandparents. Individuals wanting land had to participate in some form of local group intentionally constructed by a leader. During this period again, natural resources were similarly perceived by the state and settlers in Chittar. Accordingly, the Population Census of 1961 lists the Chittar region under the area names of 'Food Production Areas I, II, III, and IV'.

3.3 First attempts to recapture land

Following several developments during the 1960s, the state stopped its policy of settlement schemes and even changed its attitude towards settlers. The earlier atmosphere of participation and joint venture was now replaced by a more tense relationship, especially vis-à-vis those who continued to take land. They were now called 'encroachers' who needed to be 'evicted'. These encroachers – now not in a position to participate in state programmes – had to search for, or were identified by, other alliances, especially political movements.

Kerala has a long history of socio-political mass movements, including movements struggling for agrarian reform – which in turn are closely associated with the rise of the Communist Party (CPI). The party won the first elections in Kerala (1957) and tried hard to fulfil promises regarding land reform. However, they were removed from office by an alliance of opposition forces. The second elections (1960) were won by the Congress Party based on a coalition arrangement. Thus, the CPI was in opposition, and the support of the cause of the settlers (seen as neglected by the new government) became a means of active opposition politics. Many settlers thus participated in the support programmes of the left (or vice versa), leading to conflicts with the state (the episode narrated by P.K. belongs to this period too). As one strategy to reduce conflicts, the state attempted to regularise 'encroachments': all those who got, or took, their land prior to April 1957 (the year in which the previous government came to power) would receive land titles and thus be entitled to participate in other government schemes (such as agricultural extension). All others were to be evicted – and conflicts increased. Communist leaders such as A. K. Gopalan actively participated in the settlers' struggle through means such as hunger strikes. Very often, settlers succeeded in keeping their land, or to get land at other localities, to the annoyance of the Forest Department.

Forest conversion was not only continued illegally by people (who participated with those supporting them), but legally by the authorities (participating with others). In the second case study area further south (as well as in many other places), the Forest Department clear–felled natural forests to establish eucalyptus plantations, with the support of the World Bank – a new alliance which had been identified for participation.

Participation within networks of relatives was practised by others to obtain (forest) land in Chittar. Through the land reform programme, agricultural labourers received land titles (up to 10 cents) for the plot on which their hut was located. Many report that they cashed in this new asset by selling it. With the money obtained, they got more land in the frontier region of Chittar, with the assistance of relatives already residing there.

Perceptions of natural resources during this period did not differ much between settlers/encroachers, mass movements, (to a large extent) the Forest Department and foreign donors: all were interested in the land component to increase production. Differences did arise in the definition of production: settlers/encroachers were interested in subsistence and increasingly in cash crops, while the Forest Department and the World Bank were interested in raising tree crops (eucalyptus) as industrial raw material. The tribals' production interests got lost; they had nobody to participate with.

3.4 Massive conflicts

In 1975, Prime Minister Indira Gandhi declared an internal Emergency in India, so curtailing the influence of political parties and mass movements. The Forest Department was able to undertake massive evictions (in which, probably, only the police forces participated). But people in Chittar say that many of those evicted soon returned after the lifting of the Emergency in March 1977; conflicts along the forest boundary increased again, and so did alliances, pressures and arguments. There were actors within government who tried to press harder on the encroachers/settlers. Committees were established, and based on the recommendation of one, a new deadline for the regularisation of encroachments was announced (1 January 1977). However, efforts by the Forest Department to evict non-conforming settlers and encroachers often failed due to political interference. By now, politics in Kerala was dominated by two groupings each consisting of many political parties. Elections brought these coalitions alternately to power; thus, even small parties – holding the balance of power on coalitions – gained crucial influence and bargaining power for their members in government. One elderly man in Chittar proudly recounts his struggle with the forest authorities who tried to seize most of his land with the participation of the police. He succeeded in getting a stay order, and in the meantime contacted a minister (from a small party) in the state capital; as a consequence, he was not evicted.

Support for the Forest Department to prevent further encroachments and to evict encroachers gained momentum in the early 1980s. Many factors contributed to this, including a severe drought in 1982/83 (lack of monsoon rains affected not only farmers along the forest fringe, but – since Kerala depends on hydroelectric power – all people through power cuts); a political vacuum within left mass movements (the struggle for land reforms was over; mobilised youth considered the emerging environmental agenda as a new field for participation in political activism); and the Silent Valley controversy (opposition to the construction of a dam in an ecologically sensitive area). At an all-India level, a new Forest Conservation Act was enacted in 1980, giving the central government more control over forest lands. Environmentalist claims on the resource bundle got more and more support from different people and groups – who thus participated in the creation of an environmental discourse in Kerala – putting pressure on the state and supporting environmentalist circles within the Forest Department.

How did the emerging conservationist identity of the Forest Department manifest in Chittar? Here, illegal land seizures continued, though they were restricted to smaller areas along existing agricultural lands. The Forest Department's power to enforce its claims increased. In 1982, a team of Forest Department and Revenue Department officials started joint verification: The officials had to identify on the ground those settlers who came legally before, or illegally after, 1 January 1977, and had thus to define and create a forest boundary. They marked their boundary with wooden stakes, and sent reports to Trivandrum. But the processing of the files took time, and – as one forest official remarks – the stakes 'started walking'. For the next four or five years, the struggle to make the boundary final (by constructing boundary stones and walls), based on renegotiations between settlers and officials, continued. Settlers were trying to save as much land as possible, and for this purpose approached courts and tried to construct participation and networks of influence. Agitations broke out too; a little south of Chittar, 15 people were reported injured and one killed. But by 1987, the boundary wall was constructed in Chittar.

3.5   The last ten years: a period of unsettled tensions

The period since the mid-1980s is seen as a new era in which an atmosphere of forest protection and conservation dominates. My own analysis shows, though, that this new era is a rather tense one. In Chittar, almost no encroachments are reported after 1987. The boundary wall is a visual symbol of the power of the Forest Department. But the influence of the Department goes beyond the wall, into the cultivated land of the farmers. By state law, much of the agricultural land along the forest fringe still belongs to the category of Reserved Forests, land long ago declared as exclusive state property; for this land (*puduval*) one does not get a bank loan or Rubber Board support (see the example of P.K. in Appendix) and farmers are not allowed to cut trees.

The Kerala government appears ready in principle to accept that these lands are *de facto* under the management of settlers. But this time, other actors are not willing to participate. Providing land titles to lands lying *de jure* within Reserved Forests requires reservation, that is its legal exclusion from Reserved Forests. However, under the Forest Conservation Act 1980, conversion of reserved forests to other uses needs central government agreement. Thus, most requests by Kerala to legitimise forest use changes are pending, in spite of the fact that farmers have cultivated the disputed land for a long time. And so, the Forest Department continues to police the fringe. A few years ago, it introduced the so-called Forest Station System in Chittar, similar to police stations. The range officer's building consists of a small office at the front, and a huge prison cell in the rear, with iron bars from floor to ceiling.

4. Discussion: who participates, how, with whom and for what?

This brief description of past episodes of 'participation in natural resource management' along the forest fringe in southern Kerala is intended to provide insights for the discussion of present participation efforts. What are such insights? First of all, the description indicates the variety of meanings of 'people', 'participation' and 'natural resources'.

4.1 People

A rough typology (including associated connotations) of involved actors includes the following: the early settlers considered by the state today as accepted partners; the later settlers labelled by state agencies as encroachers, and seen as a threat to forests; the tribals, treated in a rather patronising manner by almost all other parties; the Forest Department's identity as producer of industrial raw materials; the Forest Department's (later) identity which is increasingly concerned with conservation and biodiversity; industries interested in forest produce, influencing government to make Forest Department produce required raw materials; environmentalists, being themselves a group of otherwise heterogeneous actors, perceiving forests as threatened by settlers/encroachers as well as the state; political movements supporting land distribution to the rural poor; and foreign donors. This (incomplete) list illustrates Oakley and Marsden's statement that 'different interest groups struggle for control of available assets and resources', and that development interventionists such as government departments or donor agencies are simply actors or interest groups among others. And last but not least, it supports their critique of the notion of community (who, in the above list, is the community?).

4.2 Participation

The description showed many forms of participation, responding to changing objectives and changing alliances, despite being concerned broadly with forest fringe management: participation in a government scheme to convert forests to agricultural uses (settlers, government); participation in NGOs to gain access to land (caste organisations, groups constructed by leaders); participation in a government scheme to convert forests to industrial uses (Forest Department, World Bank, industries), or to evict encroachers (police, Forest Department, environmental movements); participation in political networks to avoid eviction (political parties); participation in the creation of an environmental discourse (the press, activists); and so on.

4.3 Natural resources

Perceptions of natural resources such as forests can change over time. Examples include the land component for agricultural production; the land component for the production of industrial raw materials; NTFPs of importance to tribals; NTFPs of interest to settlers/encroachers (fodder for cattle, green manure, some firewood, poles); and the ecological component (such as water retention capacity).

Besides the need for differentiation of basic notions, the description suggests that participation is a complex concept that encompasses social actors' interests, their purposeful selection of partners for participation; their strategic interaction – and active non-interaction – with others; and their capacities to make claims sound attractive and just. Participation can mean very different things to different social actors, though all use the same word within a specific context.

Participation as practised by social actors becomes an enabling condition for some while being a restricting condition for others. Whether participation has an enabling or restricting influence, however, is contingent – but some social actors have better means to define participation – that is, to influence interaction with others – and thus to shape the conditions of participation more powerfully. This also indicates that the less powerful may be drawn into, or forced to, participate, or they are capable of using the 'space for manoeuvre' (Long and Long 1992) left to them as a resource to reinforce their claims.

Participation appears in this analysis as an important means to struggle for one's vision of development within wider social arenas. Participation in natural resource management is sought with those that help in achieving one's own claims on a specific component of the resource bundle. Such participation alliances between (often unequal) partners are facilitated by converging interests and supported by related discourses; and split by changing interests and the emergence of new discourses.

Finally, participation is not an ahistoric phenomenon. The past influences – but does not determine – the present. In the following sections, the insights gained from the historical analysis are used to reflect on three highly interesting participation-related developments in Kerala.

5. Present dynamics

Three recent important developments regarding participatory natural resource management in Kerala are the new *Panchayati Raj* system; the People's Campaign for Decentralised Planning; and the new World Bank-supported Kerala Forestry Project.

5.1 *Panchayati Raj*

Forests are considered to be a public resource, controlled and managed directly by the state in the name of the people. People participate at state level through democratic means, but there has been no effective democratic participation at the local level, though people's representatives were elected in *gram* (village) or district *panchayats*. More effective measures to strengthen local governments were initiated in the early 1990s and negotiation of actual procedures, and their implementation, is now an ongoing political process. This includes the definition of the range of participation local governments shall have in the context of natural resource management. Regarding forests 'on the other side of the wall' (presently controlled by Forest Department), it is understood that there are no plans to shift or share jurisdiction with the *panchayats* (except some remaining activities of earlier World Bank-supported social forestry programmes), although large areas of *panchayats* along the fringe are covered by forests.

5.2 People's Campaign for Decentralised Planning

To support the above decentralisation process, the State Planning Board in 1996 initiated (with the return to office of the Left coalition) a People's Campaign with the objective 'to empower *panchayats* and municipal bodies to draw up ... schemes within their respective areas of responsibility'. The intention is that 35–40 per cent of the Five-Year Plan's outlays consist of schemes formulated and implemented from below (Isaac and Harilal 1997: 53). The mass campaign seeks the participation of the people in each of the 1,000 or so *panchayats* in Kerala to assess resources, identify needs and develop ideas for development. As a first step, each *panchayat* published its own (and first) development report. For the present study, three such reports from *panchayats* along the forest fringe were analysed. In Chittar, reference to forests is made indirectly only. First, land titles are demanded for all the farmers having *puduval* land, and prohibitions on the felling of trees on these lands are to be revoked. Second, the document demands the fencing of forests to prevent attacks by wild animals. The same points are raised in the adjoining *panchayat*. This document even suggests that if materials for fencing are provided by the state, then farmers themselves will participate (that is, by constructing the fence). In the case study locality further south (where the Forest Department cleared forests to plant eucalyptus), the document has one chapter on forests, stating (among others) that the eucalyptus plantations created problems for the local people who needed firewood and cattle fodder from the forests. The document criticises the state departments in charge of the forests for their 'unscientific and non-discriminatory afforestation programmes', it demands that future afforestation programmes 'will be implemented only after proper study and *with the consultation and recognition of the Village Panchayat*, and that eucalyptus be replaced by local fruit trees (texts from English translations of respective documents; emphasis added).

5.3   The World Bank-aided Kerala Forestry Project

This project was officially launched in mid-1998 and consists of several components, with the participatory management of degraded fringe forest areas as the most interesting here. The project can be understood as an attempt (by the World Bank) to spread the Joint Forest Management concept (developed in other Indian states) to Kerala. Participatory management involving tribals and the population along the forest fringe is 'to develop among them a sense of stewardship for the resources. ... Forest department and [NGOs] will work with Village Forest Committees to develop microplans for rehabilitating forests in their areas, jointly determining which species to plant and how to allocate resources' (World Bank 1997; see also World Bank 1998). The project is in its initial stage and it is interesting to follow the discourses used by the involved parties in expressing project intentions and means of implementation. In the words of a World Bank staff member:

> Kerala is one of the few states in India where the Forest Department has developed a vision statement in consultation with non-governmental organisations (NGOs), community leaders, and other stakeholders. The vision statement is intended to enable the Forest Department to manage the sustainable development of forests in Kerala in a participative, people-centered way that ensures optimum land use; conserves the fragile ecosystem and protects the natural forests; meets the needs of forest communities and the needs of industries based on forest products. (D'Silva 1997: 52)

The information that emerges from forest officials is more pragmatic. Statements suggest that the Kerala Forest Department does not share the World Bank's enthusiasm for Joint Forest Management as practised in other Indian states, and is more interested in developing a 'Kerala way to participation'. It is argued that the notion of 'forest communities' has a different meaning in Kerala. Dependency on forests is limited with regard to firewood and NTFPs (except certain tribals), and the land component continues to attract most attention of the people along the fringe. A Kerala NGO has been entrusted to develop first models for such a Kerala approach. Discussions indicate that the proposal argues for proper targeting to identify those who really do depend on forests products. The proposal also seems to suggest a soft approach to promote participation using facilitators from NGOs and the Forest Department, and giving the groups of beneficiaries much room to develop (and implement) their *own* ideas. In addition, it is understood that a formal link to the emerging *panchayat* system is proposed, especially regarding financial arrangements, monitoring and control – issues apparently seen differently by the Forest Department, who seems to favour an approach including NGOs only, excluding *panchayats*. It appears that the NGO proposal does not touch upon sensitive questions such as *puduval* land.

6. Conclusion: possible futures

Three challenging developments have emerged in Kerala, and all three aim at increasing people's participation. How do they take account of the unsettled tensions of the past? And can the present analysis provide recommendations for these ventures to better consider the past?

In this assessment, the present situation along the forest fringe in southern Kerala is tense. Some argue that the key meanings guiding present actions are few, and that conservation is a powerful and widely shared value today. But the present study suggests this to be a rather static perception, and observers at the fringe state that the value of the land component continues to be important: that it is just suppressed, perhaps temporarily, by other dominant meanings. Recent initiatives intend to create an atmosphere of mutual trust and confidence among the main actors, and some are based on the assumption that this confidence exists. However, the non-inclusion of forest-related issues in the new *Panchayati Raj* set-up and the rather hostile statements in the Panchayat Development Plans are not necessarily indications of a participatory mood. And the discourse on people's participation in the World Bank document cannot hide the underlying tensions manifest in issues such as who should be allowed to participate, how and who should be excluded.

Does this assessment have any value for normative debates and thus policy recommendations and advice for development practitioners involved in facilitating and creating an enabling environment for people's participation in natural resource management? It is a difficult question; a partial answer is developed here through sketching possible futures of the new World Bank-supported forestry project. These speculative sketches are inspired by what Robert Chambers (1983) once called the two cultures of 'negative academics' (criticising projects because that's how they survive in academics) and 'positive practitioners' (highlighting the creative and positive achievements of their activities).

6.1   The 'negative academic's' view of the future

The negative academic foresees future developments as follows. A major debate will arise among the main actors on who should have a say and thus be allowed to participate, regarding what and to which degree. This debate and negotiation process (among unequal partners) also defines the space for manoeuvre each actor gets in the project. *De jure*, it is now the Forest Department that is in overall control, plus those individuals who receive legal permits to use forest products. *De facto*, though, settlers, tribals and influential outsiders are actively utilising fringe forest resources. Reading the recent history, our negative academic assumes that the Forest Department is not willing to hand over control of forests (and thus the whole bundle of resources) to others. Forests are the Department's main resource and justification for existence; a whole forest administration has developed for this purpose. And even if the Forest Department would be prepared to share responsibilities, other actors (central government, sections of the public) would not allow it.

On the other hand, measures are required to utilise forest fringe resources better because they have become degraded. To arrest further deterioration, those actors using it *de facto* need to be involved. This is a fact, because the policing function of the Forest Department has not been effective in preventing resource exploitation. Technical options exist to improve the resource base. The Forest Department, so the argument goes, is in a dilemma. World Bank assistance is attractive to the Department, which needs such support not least to meet liabilities created under the previous World Bank-aided Social Forestry Project; and the Department is interested in getting the 'institutional development component' (trainings, GIS, transport facilities) of the aid programme. To get this support, Forest Department has to accept the people's participation component which is being strongly pushed by the World Bank. But in order not to arouse forces that may challenge the power of the Department, it prefers to collaborate with (small) NGOs for the implementation of the participation component. Formal local government involvement is avoided, with the Department pointing out the political nature and unreliability of the *panchayats* (a discourse often well received by foreign donors).

Some *de facto* users will show a keen interest in improving the resource base, because they are facing increased difficulties in collecting NTFPs. They are thus ready to participate with the NGOs (selected by the Department) to make certain investments in the forest fringe. But they want these investments to be protected; they start arguing that 'they did it', and so 'want to own them'. This calls for the Forest Department to hand over considerable control to the *de facto* users who may even network (participate) with the *gram panchayats* or the People's Campaign to support these arguments. To prevent such developments the actual influence of the beneficiaries is kept low by giving experts a key role in defining fringe forest management, and the issue of land titles is not touched upon. Thus, the people's participation component in the project remains small. But the local people formally included in the Village Forest Committees will utilise these (limited) structures as resources, and will handle them strategically. They will please the officials and the visiting foreign donor representatives by following some of the formal agreements, while continuing their informal activities. Improvements in the resource base are thus minimal. As time goes by, so does the project – says our negative academic.

6.2 The 'positive practitioner's' view of the future

He/she disagrees with the negative academic and argues that the project is fully aware of underlying tensions and political dimensions of participation, maintaining that a donor-sponsored project gives the space required to innovate new procedures. He/she accepts that some donor representatives may be inclined towards a community approach, but that the Kerala Project staff (especially at the higher levels) know the game in their state, and thus also the issues of actor interests, arguments, strategic alliances and politics.

The positive practitioner argues that the definition of participation is indeed a complex issue, and that the sensitised Forest Department staff have to search for options within an arena of many expectations and pressures (from the *de facto* users of the forest fringe and their allied forces to the donor or the central government). As a consequence, the participation component has to be small – at least initially – to leave room for experimenting. Within these pilot projects, an atmosphere is created (for example, by using PRA/PLA tools) to enable discussions of all pertinent issues among the stakeholders, including questions of land ownership; and that – through processes of mediation – new solutions are, and *have to be, found* that can be accepted by all involved (he/she thus challenges the negative academic). Further efforts are then required from those in charge to strengthen participation so that all involved actors are able to test, monitor and evaluate generated ideas, and to implement flexible and lasting solutions in an atmosphere of a shared vision.

The positive practitioner is confident that through a careful working of the key project staff, involved actors will be ready to give up previously held interests, influences and powers, and are ready to learn and practise different, and new, cultures of interaction, resulting in sustainable resource management. But in case results are not forthcoming as expected, then the pilot phase would have to be extended.

Two possible futures have been identified in this chapter. To come back to the original question about whether the present analysis of the past can contribute to normative discussions to help guide future actions: perhaps it might prove its usefulness not by claiming to be in a position to provide clear policy recommendations; but rather in having a strength in critically accompanying and constructively challenging ongoing processes of participatory natural resource management. In this context it is important to recognise that practitioners and researchers too should not be perceived as members of a community within some sort of knowledge system oriented towards development, but as interest groups with conflicting views and interests – themselves having to search for new means of interaction.

Appendix 2.1: P.K.'s story

Mr. P.K. and his family live on the upper slopes at the very end of a side valley in Chittar. P.K. interrupts our discussion for a moment. Because of our arrival, he cannot distribute the mail himself, so he hands over a bundle of letters to his wife. His part-time job as assistant postman is important and needs to be done regularly. The income is small but is paid throughout the year – an important component that he is able to build into the livelihood of his family. Another component is the licence he got from the Forest Department to collect NTFPs (in front of the house, tree bark is drying in the sun). The most important asset, however, is the 1 acre of land that he owns (1 acre = 100 cents = 0.4 ha). This land borders the forest and is surrounded on three sides by a stone wall about half a metre high.

P.K. reports that his grandparents came to this locality around 50 years ago, with five other families; together they received 16 acres of forested land from the government for conversion to agricultural use. His grandparents later divided their share among their six children, with each child receiving 0.45 acres. P.K.'s father purchased the shares of two of his siblings and so managed to compile about 1.3 acres. A few years ago, he again divided this land among his three children. Thus, P.K. got 40 cents.

Today, he has 1 acre, but at one time had 3 acres (P.K. proudly states) and he starts to explain the difference. Around 1965, a rumour went through the valley, spread by a Marxist Party, that people were allowed to convert more forest land. Many local party members used the opportunity and started to clear the forest, including in P.K's neighbourhood. They organised collectively and started to plant hill paddy, sharing the harvest among themselves. P.K.'s family did not join the group, but also occupied 2 acres of forest, just adjacent to their land. P.K. says that they *had* to take this land, to secure it, to make it clear to the Marxists that this piece of land is 'reserved', is 'ours'.

In the second year, hill paddy had already been sown when the Forest Department intervened. The settlers were evicted, the crop was destroyed and the land immediately planted with teak. P.K.'s family, however, succeeded in keeping the new land, because they were able (so P.K. claims) to convince the officials that they were not Marxist group members. This success gave them confidence, and they started to cultivate the additional 2 acres in a more permanent manner. They invested a lot in terracing the steep land and were happily looking forward to the first tapping of rubber trees which would have been possible around 1985.

But things turned out differently (and P.K. gets more excited while recounting these events). One morning, a high-ranking Forest Department official arrived with a group of workers. He declared that P.K.'s agricultural use of the 2 acres was illegal; that the land belonged to the government. The official set out wooden stakes, and the workers started to construct a wall along the emerging boundary line. On the other side of the wall, they cut down all rubber trees, uprooted pepper plants and started planting bamboo. Only 60 cents remained with P.K. from the original 2 acres – a very bitter time.

That's why he has 1 acre today: 40 cents of land with land titles and 60 cents without (*puduval*), left to him for cultivation by the Forest Department, but still considered by them as their property – land against which he does not get loans and for which he is not entitled to agricultural extension advise or the attractive packages offered by the Rubber Board.

Note

1. This chapter emerged from ongoing research into the social production of the forest boundary in southern Kerala. Methodologically, the research consists of two case studies (Chittar and Peringamala), interviews with resource personnel and involved actors at different levels, and the consultation of unpublished secondary sources (grey literature). The research is funded by the Swiss National Science Foundation's Priority Programme Environment (SPPU Module 7), the Swiss Development Cooperation (SDC) and the University of Zurich. Many thanks to Bhaskar Vira and Heidi Stutz for comments on an earlier draft.

3
Constructing the Future: Dynamics of Local and External Views Regarding Community-based Resource Management

Diny van Est and Gerard Persoon

1. Introduction

In the global sustainability debate the emphasis is on the future, the long-term future. Unlike anthropologists, environmental scientists and conservationists are obsessed with the future: scenarios and models are their most important instruments, environmental concepts (sustainability, regeneration) and policies (including conservation, restoration of ecology) are outcome- and future-oriented. This future orientation has two faces: first, visions of an apocalyptic future are used to justify conservation interventions (Western 1994; Leach and Mearns 1996; McNeeley 1996); second, images of a wanted and preferably better 'green' future can confine opportunities for innovation and change. Signs of hope (often inspired by non-Western cultures) are urgently needed to generate sufficient support for these alternative visions. All kinds of models, scenarios and policy instruments are being developed to turn these alternatives into reality.

Recently, conservationists have found natural allies in local or indigenous people. Increasingly, these local people have been defined as partners in nature conservation, because they have been the guardians of the resources for extended periods of time. Representatives of these communities or their activists and academic spokesmen argue that this cooperation should be based on the assumption that local people's needs, phrased in terms of conditions for survival now and tomorrow, are to be met first. Rights of access to these resources are based on historical reasoning, as they were always the rightful owners of the resources in the past before being deprived of them by external forces.

The claims are often combined with the present-day rhetoric of sustainability. Also, in retrospect, local groups consider themselves as the guardians of their environment. These claims are made to ensure their future involvement.

Although conservation is explicitly addressed to the long-term future, the development bureaucracy (in its plural manifestation) has different time perspectives. Its mindset is narrowly structured in rhythms of short project cycles governed by new idioms and fashions. In addition, this bureaucracy is in many ways closely linked to the rise and fall of politicians or political parties. This also holds for bureaucratic institutions that are engaged in nature conservation. They have adopted and imitated the style of development agencies in general.

In this chapter we want to explore how, in the 'real world' of natural resource management, these various time perspectives or visions of environmental futures come together and interact. We try to understand this interaction by making the temporal aspects more explicit for two cases. The first case is taken from the Philippines, where an indigenous people are granted access to ancestral land on a collective basis but with an obligation to manage the area in a sustainable manner. The second case refers to a national park in the north of Cameroon where an ethnically highly diversified 'local population' are urged to participate in the management of wildlife. Both cases bring together various groups of local people, NGOs, national and even international agencies who all operate from highly varied 'time scapes' (Adam 1998). Before introducing the cases, however, we will discuss the concept of the future in anthropology.

2. The future

Generally speaking the future is conspicuously absent as an object for research in the anthropological literature (Wallman 1992; van Dijk 1997). In the field, anthropologists are usually more interested in the present and its genesis. In other words, they stand with their back to the future. Even in those studies that deal explicitly with the anthropology of time, the future is thought to be of only minor importance (for example, Gell 1996: 314). It appears that the future is not considered a serious issue for investigation during fieldwork. Research handbooks also pay little attention to this topic. It is only once anthropologists return from the field that they start to reflect on the 'future'. They often feel the need to finish their monograph on a particular people with a chapter that includes 'the future' in its title. Usually this chapter is a reflection on what might happen with the people in the future. This is not based on field research, nor does it reflect how local people look at the future. It is, however, based on projections of what the anthropologist thinks might happen. These projections often prove not to come close to reality (see, for example, Boissevain 1992). In such work it is not the native's point of view on the future that matters, since it is the anthropologist who is looking ahead. In doing so he/she brings in all kinds of external factors and projections. There are numerous examples of monographs in which the future is dealt with in this way (exceptions are the studies that deal with messianic movements or with revitalisation movements, mainly originating from the Melanesian region).

A simple argument for anthropologists in general being more past- and present-oriented than future-oriented is, of course, that there is no empirical evidence for these images of the future. But this does not automatically imply that ideas about the future among other people could not be studied as a serious subject. We are convinced that people in general are much more occupied with the future in their daily operations than with the past. People plough and plant today in order to harvest towards the end of the season. They invest in boats and nets in order to catch fish in the near future. There are also other kinds of future which imply alternative options, not only in the field of natural resource exploitation but also in social domains, in relation to kin as well as to those in power. People may also plan for the more distant future, for themselves or for their offspring. They decide on the resources needed to make particular social or economic investments. They may do so individually or collectively.

Particularly in rapidly changing circumstances, neither the past nor the present can teach us everything about the future. Even the present can be understood only as soon as we understand the image or perception of the future as held by the people involved. People may opt for radically different alternatives. By focusing too much on the present and the past, researchers miss out on innovative possibilities and options for change. For instance, some remarkable examples emerge from the study of agricultural transition. People make changes, organise their lives differently, value familiar ways of behaviour in a new perspective and take different directions because of changing circumstances. What for some seemed an unrealistic distant future might be made more accessible through innovative and risk-taking individuals (see, for example, Conelly 1992).

In more general terms it may be useful to differentiate between different kinds of futures. In the first place there is what Bourdieu (1963) labelled the 'forthcoming future', the concrete horizon of the present. It is the kind of future that is expected and that can be influenced or manipulated to a great extent. In the second place there is the unknown or maybe even unknowable future, which people may also look upon as not to be manipulated, 'in the hands of God', or dominated by supernatural forces.

3. 'Community' and 'management' in community-based management

In the concept of community-based management the terms 'community' and 'management' are often used in a rather loose way. Moreover, it is not always clear what these terms actually mean in the field. What kind of phenomena or images are covered by these concepts, and what do they mean with respect to the actions of real men and women and in relation to particular resources, changing in time and space?

3.1   Community

Terms such as 'community' or 'local people' are often used (in the literature on natural resource management) to imply *'that a community can be treated as a single entity with a single set of interests'* (Crehan 1997: 227). Reality is, however, far more complicated. Many authors (Benda-Beckmann and Brouwer 1992; Gibson and Koonz 1998; Leach 1997a) emphasise that this 'community as a whole' concept is highly problematic. They state that one has to reckon with realities such as local power dynamics, differences in access to resources between men and women, several ethnic groups, age groups and individuals. The fact that ideals about communities differ from reality has important practical implications for resource management in general and for the specific cases we present in this chapter. This takes us to the second term in community-based management.

3.2   Management

In the sustainability debate we see that the term 'resource management' differs from one interest group to another. Interest groups may range from international to national and local levels. Management is also used in a variety of empirical situations ranging from controlled utilisation, protection and maintenance to purposeful regeneration. It also includes the transition from one type to the other (Wiersum 1996). Differences between professional nature conservation organisations and local people are apparent in terms of objectives, views and practices. The first group wants to protect 'ecosystems' (forests, wet-lands and savannah) as a whole, while local people are more interested in specific resources relevant for fulfilling human needs. In the context of this discussion we would like to take a closer look at a Philippine case and a Cameroonian case in which the aforementioned levels inter-act within projects.

4. The Bugkalots' rights to forest resources (Philippines)

In order to understand the situation of indigenous people such as the Bugkalots in the Philippines it is necessary to provide some back-ground information. As in many other countries the indigenous peo-ples in the Philippines lost their lands to colonial governments over the course of history. It is estimated that about 10 million people belong to the country's indigenous peoples. Most of them live in and around classified forest areas. Although the occupation of that land is illegal (it is reserved for logging concessions, and nature and watershed protection) there are a number of ways through which they try to secure their use rights.

Soon after the 1986, revolution which brought Corazon Aquino to power, the indigenous peoples supported by advocacy groups in which scientists, church leaders and journalists were active, successfully lob-bied the Constitutional Commission for recognition of their ancestral domain under the new Constitution. Pushed internally by increasing pressure of indigenous groups and their advocates and externally by the globalisation of the indigenous movement in which the Philip-pines has played a prominent role, the Congress finally drafted an Ancestral Domain bill towards the end of the Aquino administration but the bill was not endorsed and its momentum was lost until the end of 1997.

The Department of Environment and Natural Resources has designed and implemented a number of programmes which, over the last couple of years, have been based on the idea of what is being called community-based management. These programmes were launched under the slogan: *People First, Sustainable Forest Will Follow.* These programmes, funded by the Asian Development Bank and the Japanese government, gave individual forest farmers or 'communities of farmers' Certificates of Stewardship over a piece of forest land for a renewable period of 25 years. Many of these programmes are based on the idea of granting access to resources under particular conditions (e.g. reforestation and sustainable use) and for a fixed number of years (Lynch and Talbott

1996: 86–9; Pasicolan 1996). It is important to note here that the land remains the property of the state.

4.1   CADC principles

In 1993 the Department of Environment and Natural Resources (DENR) issued an administrative order (no. 2) on the *Rules and Regulations for the Identification, Delineation and Recognition of Ancestral Land and Domain Claims*. This initiative was probably taken to follow up on the discussion in Congress on the ancestral domain bill, which was not pushed through because the attention of the Congress was shifted to other subjects. Through this order DENR, the department that is entitled to exercise exclusive jurisdiction on the management and disposition of all lands of the public domain, aimed to find a way out of the continuous conflicts with local tribes and communities over the land, the forests and the other resources.

Ancestral domain and land claims are now found to be necessary for access to critical watersheds, particularly for domestic water use, wildlife sanctuaries, wilderness, forest cover or reforestation, as determined by appropriate agencies with the full participation of the indigenous cultural communities (ICC). Numerous indigenous groups have applied for these certificates. Even though the certificates did not give them full ownership of the land, receiving recognition of a legal claim was already a major step forward. Until late 1998 almost 200 claims were granted, covering a total area of more than 2.5 million hectares of many forest lands. Following the establishment of the logging ban or logging moratorium in many areas in 1992 and 1993, the issuing of these Certificates of Ancestral Domain Claims (CADCs) seemed an attractive solution for DENR to allow for more local involvement in natural resources management in accordance with the spirit of the process of decentralisation and finally addressed the long neglected claims of the indigenous cultural communities in the country.

4.2   The CADC of the Bugkalot[1]

The Bugkalot people in the municipality of Nagtipunan (Quirino Province) was the first indigenous cultural community which was a beneficiary under the DENR Administrative Order. In June 1994 the community was granted a CADC through which its members were collectively granted land/resource tenure security over 108,360 hectares of their ancestral domain. This area covers about one third of the tribe's aspiration to acquire control of their entire ancestral domain, which spans three provinces. After the issuance of the CADC in June 1994, the Bugkalot focused their efforts on the preparation of a resource management plan for which they obtained USAID-funded technical assistance from the Policy Studies Office of the DENR (DENR 1994). The management plan, including a resource inventory, was discussed a number of times with the Bugkalot leaders and governmental agencies involved in the area. Finally, in 1995, the plan was approved and it now constitutes the framework within which more detailed plans can be evaluated.

4.2.1 Matters of time

The history of the Bugkalot (a 'group' of about 13,500 people) in relation to their environment over the last decades, can largely be written in terms of loss of land and forests resources. This loss has to be attributed to encroaching farmers and logging companies. Though the Bugkalot are known to have been fierce head-hunters in the past, a reputation which they cherish, they have not been able to resist these strangers in their homelands. The awareness of these historical facts has a profound influence on the way the Bugkalot view present and future resource exploitation.

The area, which is now declared an ancestral domain, covers 12 villages (*barangay*) with an estimated population of 4,500 people. In a sense the idea of ancestral domain is a return to the situation before 1960. Until then land had been Bugkalot domain in practice: they could clear gardens and hunt at will. During the period of cultivation the land and its production were more or less private, but once the land was left fallow it gradually became common property again. One should, however, not romanticise their environmental use: there was little, if any, actual management exercised over the closed canopy and secondary forests.

The CADC is in itself an interesting manifestation of different time perspectives. On the one hand, it is an expression of recognition of history and rights that can be derived from historical processes. There is certainly an aspect of compensation (*Wiedergutmachung*) for all that was done to the Bugkalot and to the land that they have occupied since 'countless generations'. On the other hand, resource management deals, by definition, with the future, particularly the future state of resources (expressed in terms of biodiversity and productivity). The combination is partly based on the assumption that the indigenous people will still practise some of the basic principles that they employed in resource use in the past.

It is only in relation to the CADC, and in particular during its preparatory phase, that the Bugkalot could actually start thinking about the future use of their resources partly under their own management. But the idiom in which they think about the future is not so much derived from their traditional use, which was part of the justification for granting them the rights, but from more recent forms of exploitation. Modes of use of resources as employed by encroaching outsiders (loggers and migrants) now also determine how the Bugkalot envisage the future exploitation of their resources. Commercial logging and conversion of forest land into permanent agricultural plots are the dominant themes in their thinking.

4.2.2 Matters of management

There can be no doubt about the discrepancy between the management aims of the national government and those of the Bugkalot. The national government, strongly supported by international rainforest protection programmes, aims to maintain and restore the region's biodiversity and ecological functions, allowing only for a safely defined sustainable yield in terms of timber and other forest products. This aim is formulated after decades of heavy logging activities and slash and burn agriculture. The management aims for the area are changing simultaneously with the management structure. Management aims, as imposed upon the communities, have in practice changed from a kind of 'mining' of forest resources to aims phrased in terms of biodiversity conservation, sustainable (restricted) harvesting and 'traditional use'.

The Bugkalot look upon the severe restrictions with respect to the allowable harvest with mixed feelings. They feel that the restrictions regarding the modes of exploitation (no heavy equipment, no machines) are unfair and outdated. They themselves feel in no way bound to the modes of exploitation of their ancestors. Being indigenous is primarily used as a way to obtain rights: it does not provide guidelines for present-day or future modes of resource use.

4.2.3 Matters of community

By definition the binding element behind the CADC is the common ancestry of an indigenous people. The Bugkalot, however, never manifest themselves as a corporate social group. They have no form of social organisation that could be charged with managing the full size of their ancestral land. Ironically, it was an outside organisation, the New Tribes' Mission, that established the social unit, which successfully applied for the CADC. The non-Bugkalot missionary acts as the chairman of the CADC community, which constitutes only part of the Bugkalot people. Because of the inaccessibility of the area, internal communication about matters related to the CADC is limited. In practice only a few people are involved in the actual decision-making and in the distribution of the benefits to the 'community'.

But there are also other 'community' levels involved. The villages and hamlets within the CADC area are ethnically not homogeneous. Migrants from other ethnic groups may also live in these settlements and are part of the local communities, which interact on a day-to-day basis. Instead of supporting the sense of belonging within the Bugkalot as an indigenous people based on common ancestry, the CADC has reinforced new divisions among them. New boundaries have been drawn among the Bugkalot through newly introduced religions (variants of Protestantism and Roman Catholicism) and between them and other ethnic groups. The CADC articulates these boundaries by adding rights to natural resources to a particular group of people (the New Tribes' Mission converts). It is unclear what the outcome of this process will be.

5. People and the Waza National Park (north Cameroon)

The Waza National Park (170,000 ha) is situated in the Logone floodplain in the extreme north of Cameroon. In the past this wetland region of about 8,000 sq. km provided a living for thousands of people. The livelihood of the sedentary and nomadic population depended on the fertile and yearly flooded lands, which determined the broad range of favourable opportunities for exploitation. Moreover, the biodiversity of the area is of national and international significance. The Waza National Park supports a large and varied stock of wildlife and birds in the region.

Since 1979 the Logone floodplain has been affected by the building of the Maga dam as part of a nationally and internationally supported project for large-scale rice irrigation.The construction of this dam, combined with a prolonged drought that affected the whole Sahel, prevented flooding over a large part of the plain. This flood reduction has disrupted the living conditions of people and wild animals. In fact, two competing government policies, one promoting economic growth (rice irrigation) and one aimed at conserving wildlife through protected areas (for example, Waza Park), clash dramatically in the Logone floodplain.

5.1   WLP project: linking people and Waza National Park

The Waza National Park, now a biosphere reserve, was one of the reserves created by the colonial authorities, which became a national park during the first years of independence. The establishment of national parks like Waza has certainly contributed to the fact that a variety of wildlife is still present in this part of Africa. Serious criticism, however, has arisen with respect to the top-down way in which this park has been created (villages in the protected area were dislodged and resettled), and is being managed. As in many other protected areas in Africa (IIED 1994), this centralised management is in a crisis. Apart from the lack of inundations, which has decreased the diversity of the habitat, the park is threatened by conflicts between people and park management, especially due to poaching and other forms of illegal use (Drijver 1990: 533). Under the label of 'participation', the Waza Logone Project is trying to overcome these conflicts.

Alternative strategies for the management of Waza are now being canvassed. Among these is community-based management, whereby local participation is an important strategy for the future protection of the park, in exchange for the sustainable use of defined resources (collection of arabic gum, thatch and dead fuelwood, fishing, guinea fowl harvesting). This community management is currently in its infancy, but has been agreed upon by the government, and the next important step will be the establishment of the management committee. The project staff wishes to appoint delegates from the sedentary population, from nomads and they also want to include women and youngsters in this committee (IUCN 1997b).

If we look at the negotiated management plan the main orientations are: anti-poaching (through the strengthening and expanding of enforcement); technical and ecological management (fire management, digging waterholes, road construction, wildlife monitoring); integration of conservation in development of the region (conditional use of the park in exchange for environmentally sound behaviour as defined in the plan, and compensation for losses and damages). Such a plan allows us to discuss issues relating to a variety of time scales, ideas on management of natural resources and communities.

5.1.1   Matters of time

The emphasis on 'long-term' sustainability of the park and its surroundings raises a lot of issues on the 'cultural construction of the future' by the various actors involved. The IUCN and WWF nature conservation views presuppose an almost biological time scale: a sustainable never-ending world allowing for processes of natural evolution. The staff of the project (which has a short cycle) present the project as Development with a capital D. In this short period they want impressive results that will continue after the project has finished. But this supra-local time scape meets with a diversity of other time scales in resource use, ranging from time immemorial (Kotoko's use of fishing grounds), to temporary use of grazing lands or seasonal rights to fish. The so-called 'beneficiaries' of the project seemed to be immersed in nostalgia. The Kotoko authorities and the elders seem to yearn for the authenticity of the past. Nostalgia, however, can become a basic element in how groups deal with their present predicament and how certain claims of power and interests are substantiated in the future (see also Battaglia 1995; Spierenburg 1997). The new 'big men' (the traders), the young, the non-Kotoko and probably also women hope to gain something from the future in their contest for access to resources and their management.

5.1.2 Matters of management

The Waza Logone Project[2] is promoted as one of the first representatives of a new generation of environmental projects: linking biodiversity and sustainable resource use by people.[3] As such, it reflects changes in conservation and development thinking in general, in the approach of IUCN and WWF, as well as changes in Cameroonian governmental policy. Since 'Rio' and the SAP of the World Bank, governmental commitment for conservation (which diminished during the 1980s) has increased and resulted in the creation of the Ministry of Environment and Forest. A new law passed in 1994 facilitates participatory management.

The ambivalent role of the Cameroonian state and the experience of previous 'development' projects (rice project and Waza Park) has made people very anxious about new interventions that might restrict their lives. The Waza Logone project, although presented as new and future-oriented, is in fact repairing the problems that resulted from the aforementioned projects and is confronted with the fact that their renewed interest in local management in this specific context is rather sour. The Waza Logone Project is likely to take part in the redefinition of the access to resources that is actually taking place in the plain. The project risks becoming another part of the cake that might be appropriated by traditional and new leaders for their own benefit. When the project fails to take into account unequal access to resources and ongoing conflict over these, it risks venturing into a hornet's nest.

5.1.3 Matters of community

The project donors, WWF and IUCN, have a rather instrumental view on local communities: people are needed to manage a protected area successfully. We think this holds for the state as well. The Dutch development organisations DGIS and SNV contribute to the project as well and emphasise equity. Their views come together in the management plan.

In the management plan many equivalents of the word 'local population' are used: 'local users of resources', 'the people', 'the villages' and 'rural communities'. These neutral terms used in the report obscure the fact that in this area competing interests exist (sometimes even violent conflicts) between and within the four ethnic groups living here, between rich 'absentee cattle holders' and small cattle holders, between elite and landless, between sedentary and nomadic people, between authorities and powerless, young and old, and male and female. By using these neutral terms they produce images of communities which do not reflect the complex reality in which people interact with each other and with the environment.

The Logone floodplain is considered the land of the Kotoko people, who have lived there since time immemorial and whose authorities used to manage the resources (Lebeuf 1969). It should be borne in mind that the shift in the power balance between Kotoko leaders and their opponents influences politics and natural resource management. The management of local resources in the plain has become a battlefield with competing interests between Kotoko and Mousgoum fishermen (Van Est and Noorduyn 1997), and rivalries between Arab Shoa and Kotoko (Socpa 1992) amongst many others. Democratisation has become a strong metaphor in the bush. For the opponents of the Kotoko, especially young men who refer to their Cameroonian citizenship in this context, democratisation has become the equivalent of 'getting rid of the dominant (Kotoko) authorities' and their control over resources. In this highly diverse landscape it is therefore important to analyse which groups (ethnic, villages, political factions amongst others) or institutions are formed and constructed around issues of political relevance including access to natural resources. The 'community' might be hard to find.

6. Conclusions

In this chapter, we have tried to explore the meaning of different time scapes, and perspectives on community and management in community-based management. In this concluding section we want to compare the two cases with respect to these concepts. The term 'time scapes' emphasises the rhythms, timing and tempos of past and present activities and the interactions of organisms and matter, including their changes and contingencies (Adam 1998: 11). Placed within the context of nature conservation projects it is clear that nature conservationists, representatives of local people, donor agencies and bureaucrats operate from highly varied time scapes. However, they have to decide on a kind of middle ground in the design of projects in which power plays an inherent part. Basically, the projects are all future-oriented and incorporate future interests. Then, during the implementation phase, the divergent time scapes become evident once more, though often in a hidden manner. The outcome of this process of interaction often allows for a variety of interpretations. These interpretations constitute present-day realities, which are a combination of the planned and unintended outcomes of past actions, as well as the result of activities designed to bring about another kind of future reality.

Both cases have shown a highly diverse and 'fractured' landscape of communities. Due to the heterogeneity of communities, the 'whole community' might not always be the most appropriate unit for management of resources for two reasons. First, issues of access to and control over natural resources almost always signal arenas of competition and potential conflict in which some groups are likely to gain or lose more than others. Both cases show that interventions can make life more problematic for people, especially when interventions are related to increased income, access and power. Second, there is always the need to delineate who or which categories of people are the legitimate representatives of the local population. In that respect the 'indigenous people' label, dominant in the Philippine case though absent in the Cameroonian one, can complicate this because it might neglect community dynamics and draw dangerous boundaries in multi ethnic regions.

An important issue is the nature of the resource that is actually managed. Different resources require different kinds of management regime. Highly mobile stocks of fish or marine mammals with territories far beyond the jurisdiction of particular communities cannot be managed in the same manner as some localised non-timber forest products or slow-moving animals. Resources require specific knowledge; they are subject to specific management aims, and due to differential regenerative capacities they also require a different time perspective in order to be managed in an adequate manner. In successful management projects ecological realities should be linked with social and cultural facts of life.

We have argued here that the future constructions of 'global environmental organisations, the time scales of the development bureaucracy and the future perspectives of local people need to be examined and compared, and made more explicit. We need to develop tools and methods to study the future perspectives on these various levels, as this aspect is lacking in present work. Our interest in these future perspectives is not, to quote Wallman (1992: 2), in prediction but we are interested in the causes and consequences of images of the future. The analysis in this chapter suggests that participatory natural resource management should be accompanied by more appropriate statements about the nature of management by asking explicit questions about what (elements of the ecosystem) are to be managed, by whom (local people, indigenous people, conservationists, governments) and for whom (our generations, future generations in general or for our ancestors) and within what frames of time and space (Thomas and Adams 1997).

Notes

1. This paragraph on the Bugkalot is based on Dante Aquino's paper on *The Bugkalots of Northeastern Luzon (Philippines): Adaptation and Local Resource Management* (1996) and some field visits by one of the authors.
2. The project is mainly financed by the Dutch government, IUCN and WWF. CML and SNV support the project technically and scientifically.
3. Sub-objectives of the project are: the restoration of the floodplain by hydrological and ecological rehabilitation, to safeguard the national parks in the area, to develop resource management, capacity building, developing eco-development activities (Scholte *et al.* 1996).

Part II
Constructing Participation

4
Defining Institutions for Collaborative Mangrove Management: A Case Study from Tanga, Tanzania

Mike Nurse and John Kabamba

1. Introduction[1]

This chapter describes a case study of collaborative management from two of the coastal villages in Tanga Region, Kipumbwi and Sange. The villages (altogether some 318 households) are adjacent to a mangrove forest of 416 hectares, which provides forest products and environmental services that are a vital component of the subsistence livelihood systems of the villagers. The context for the collaborative management of mangroves at Kipumbwi and Sange was investigated by a multidisciplinary team of government staff. The work is part of an ongoing programme of technical assistance by IUCN, the World Conservation Union.

Using a variety of participatory extension techniques, the indigenous systems of management and use of the mangrove forest were analysed. Institutions were identified for the primary role of protection and prudent use of the forest. These now have an organisational basis as the Lands and Environment Committees of Sange and Kipumbwi. Equity in decision-making, and in representation within the new institutions, was examined in particular. This chapter discusses the potential for such new, externally sponsored institutions to manage the mangrove resources. The initiative has resulted in a draft collaborative Management Plan for the management of the mangroves. The Management Plan includes an Action Plan which details the activities that the villagers will carry out to achieve their management objectives. The Action Plan also contains indicators for monitoring the effectiveness of the Lands and Environment Committees.

The study found the new institutions equitable and representative of the two communities concerned. The nature of organisations and decision-making at the village was found to be evolving very rapidly in response to the initiatives of the programme, and to the national move towards more democratic processes in government. This suggests that social capital in the form of institutions and organisations for collaborative management can be heavily externally sponsored, but a great deal more effort is required to develop and check socio-economic indicators of institutional robustness. Much effort is required in monitoring by external agents – in this case, government extension staff – and by the resource users themselves, once an institution has been identified. It is much easier to build on a successful indigenous institution, as functioning organisational and resource management systems may already be in place. The organisations at Kipumbwi and Sange, though, appear at this stage to offer promise for circumstances where indigenous institutions are weak or dysfunctional.

2. A conceptual introduction to institutions and organisations for natural resource management

Rural villagers in many parts of the world share access and use of forest resources on what legally may be public land. These common pool resources (also sometimes called common property, though this tends to confound the tenure issue) are those for which use and management are shared by a community of forest resource users. There has been a popular belief that rural people are unwilling or unable to look after such forest resources properly, and that in a situation where the human population is increasing and the forests are owned by nobody, then everyone will take what they need. This would lead to the inevitable destruction of the resource. This theory of the 'Tragedy of the Commons' was proposed by the biologist Hardin (1968, quoted in Ostrom 1990), but has subsequently been hotly disputed (Gilmour and Fisher 1991).

In fact, in much of the developing world, common property provides a complex system of norms and conventions for regulating individual rights to use forest resources. Use can be defined anywhere on a scale from open access (no rules of access or use, as in the Hardin theory) to sophisticated (silviculturally) and robust (institutionally) community-based management systems. The situation is also dynamic. Management systems for common pool resources can break down and become open access or can become more sophisticated, as a result of internal and external influences. It is also important to note that often what appears to the outside observer to be open access may involve tacit cooperation by individual users according to a complex set of rules specifying rights of access and use (Nurse in press). These management systems, if generated entirely within a community, are called indigenous management systems; if such systems are assisted by outside agencies (for example, by government or project advisers), they are called externally sponsored systems (Fisher 1991).

Collaborative management refers to the partnership of local communities of forest users (almost always an identifiable group) with government in the management of a public resource. This partnership ideally takes the form of control and management of forest resources by rural people, who use government staff as advisers, rather than as protection and enforcement agents. In practice, there is a continuum of examples of various forms of participation from total manipulation to total citizenship control (Bass *et al.* 1995, quoted in Hobley 1996). The usual result of these partnership arrangements is some form of sponsored forest management system.

Collaborative management has evolved as a response to the realisation that government services are not effectively able to ensure the ecological and productive integrity of a widely fragmented forest estate through protection and enforcement alone. There is also increasing recognition that villagers using a particular forest on a day-to-day basis do have the commitment and ability to work in partnership with government in the management of that resource. In the future, by managing resources in this way, local forest users can also ensure that extractive use considers the social and environmental costs locally, in a more effective way than perhaps the establishment of royalties or stumpage fees (Bates 1995).

Most collaborative management initiatives work through a local forest user group, as an organisation that represents the institutional arrangements for management of particular forest resource. This organisation often has a committee structure. Within the bounds of common property management and use of resources, however, there are organisations and/or institutions that take responsibility and authority for management. In his discussion of local organisations for community forestry, Fisher suggests that 'the existence of effective local organisation is essential to the success of collaborative forest management programmes' (1991: 48). Fisher further distinguishes between organisations and institutions. He uses Uphoff's definition of organisations as 'structures of recognised and accepted roles' and institutions (also from Uphoff) as 'complexes of norms and behaviours that persist over time by serving collectively valued purposes'. According to Uphoff, it is possible to have 'organisations that are not institutions' (an example is a firm of lawyers), 'institutions that are not organisations' (the law), and 'organisations that are also institutions' (the courts) (1991: 50). From his experiences in Nepal with community forestry, Fisher then postulates a two-tiered model of a functioning forest management system (see Figure 4.1).

The institutional base is considered to be an essential component of the system, but the organisational superstructure not essential. The basis for this argument is that in south Asia (particularly Nepal) there exist a plethora of indigenous forest management systems, many of them recently established as a result of decline in one or a number of key forest resources (cf. Gilmour 1990). These indigenous forest management systems are commonly institutions with no organisational superstructure. Forest users may understand and have collectively embedded norms (for example, that no user may cut green products from the forest) but not meet regularly to discuss management issues or have formed a recognisable group for decision-making.

Fisher outlines the danger in external organisations sponsoring an organisational superstructure (such as a committee) that may have no institutional basis. A further hazard is in creating an additional organisation that competes or conflicts with an existing one, because of an inaccurate or incomplete investigation of the existing local arrangements.

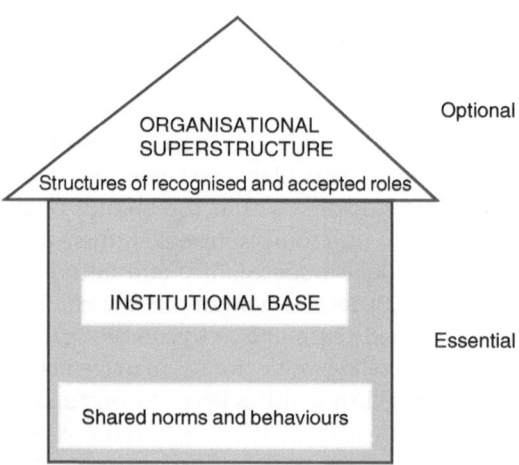

Figure 4.1   Organisational model of a forest management system
Source: Fisher (1991).

We will argue, based on the presentation of case study material, that this model is equally applicable in the Tanzanian context.

3. The practice of collaborative forest management in a Tanzanian context

Collaborative forest management finds its origin in the Asian (particularly south Asian) realm. The approach (action research[2]) and the tools (PRA and other tools for participatory assessment) are appropriate to Tanzania, but there are also a number of important differences between the Tanzanian and south Asian context. The differences which have particular relevance to the institutional aspects of resource management are as follows:

1. *Resource scarcity is not an issue which is leading to the creation of indigenous management systems.* Gilmour (1990) proposed a resource availability model for Nepal, postulating that forest users will not invest time in meetings with other users to regulate use (i.e. establish indigenous management systems) unless one or more products become scarce. Scarcity in the Nepalese context means that villagers must travel more than one full day to gather one head load of forest products. In Tanzania resources are becoming seriously depleted, but indigenous management systems are only weakly developed and often heavily reliant on traditional (tribal or clan-based) systems. The KiSa users in this Tanzanian case study reported that they felt 'powerless to intervene' when traders were cutting their mangroves as no systems were in place to protect the resources outside the sanctuary sites.

2. *There is often a more recent pattern of settlement.* The villagisation process in Tanzania (*Ujaama*) resulted in the translocation of many villagers, in a deliberate attempt to integrate the population and break up tribal groups. Whilst this generally succeeded, systems of indigenous management have broken down with the movement and communities do not always feel cohesive. One of the villages in coastal Tanzania used in this case study, Sange village, is a recent settlement. Many new villages were formed on the coast in the 1960s in response to the villagisation process because of the need for labour for sisal plantation work. Sange has experienced periodic conflict over administrative boundaries with its immediate neighbour Kipumbwi, a much older and more established settlement. The two villages have recently joined in partnership for natural resource management largely as a result of the external programme initiatives.

3. *Forest resources are often more extensive, resulting in larger, more extensive groups of forest users.* The concept of a forest user group as used in the south Asian context (particularly Nepal) does not have such obvious meaning when defining large, widely scattered groups of forest users, that do not normally meet together in a decision making forum. The response, for example, in Cameroon (Nurse *et al.* 1994), Uganda (Scott, in preparation) and Tanzania, has been to establish an organisation to represent the forest users, a committee. Committees are also common in forest management systems in south Asia, but the African examples cited tend to represent larger groups and have been given more authority. This makes them particularly susceptible to political manipulation and/or potential inequities in decision-making.

These differences provide special challenges, in identifying the institutional basis for collaborative management and in ensuring that any organisational basis is representative of the forest users. The result tends to be heavily sponsored institutions and organisations that have representation through committees, partly because of the size of the group of resource users (meetings of all users to discuss fine points of management are uneconomic and unwieldy).

4. A case study from Tanga, Tanzania

Tanzania offers an interesting opportunity to analyse the institutional aspects of community-based natural resource management, because of its cultural context and recent political history. Following a period of enforced migration and one-party rule under a socialist model, indigenous management systems for common pool resources broke down with the formalisation of power in village government. Only vestiges of these traditional institutions remain. A recent move towards multi-party democracy now offers an opportunity to build partnerships for conservation and development with institutions other than government at the village level. The identification of appropriate institutions is an interesting and challenging exercise, given the inherent weaknesses of indigenous structures for communal decision-making.

Tanga is one of the northern regions of Tanzania. Its coastline covers approximately 130 km from the Kenyan border in the north, to Sadani Game reserve in the south. Tanga's coast has one municipality, one small town and about 87 villages. About 150,000 people live in coastal villages and rely on a number of activities to maintain the household

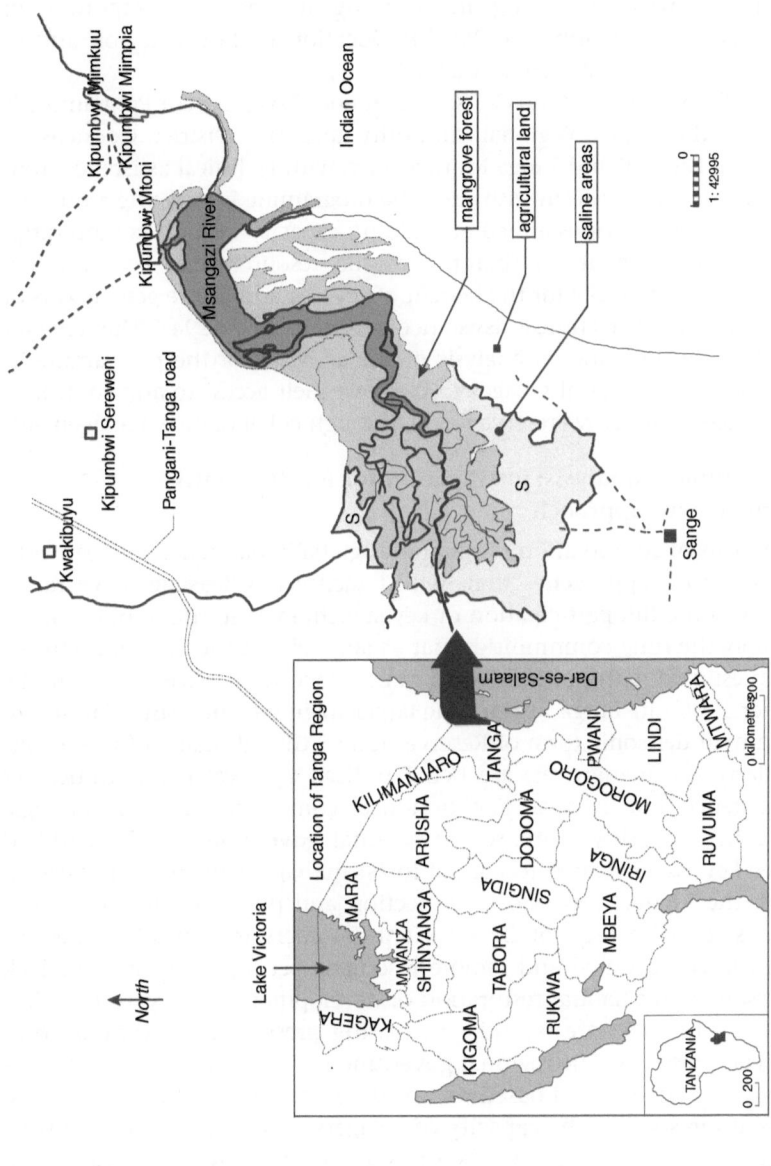

Map 4.1 The location of villages and mangrove forest in the study area

economy. Artisanal fishing and farming are the most important of these activities (Gorman 1995). The location of Tanga Region and of the study area are shown as Map 4.1.

The Tanga Coastal Zone Conservation and Development Programme is run by the Tanga Regional Authority and the District Councils of Muheza, Pangani and Tanga Municipality, with technical assistance from IUCN and funding from Irish Aid. The programme is targeting a number of integrated activities aimed at building capacity in local communities and in government, so that the coastal resources can be protected, utilised and managed for the benefit of present and future generations of residents. IUCN technical assistance began in mid-1994.[3] This chapter presents an institutional analysis of the activities of the programme in support of two coastal villages to improve their access to and control of an adjacent mangrove reserve forest, through collaborative management.

4.1   Problem analysis: justification for a collaborative management approach

The focus of conservation programmes globally has generally been concentrated on approaches that, whilst technically feasible, have rarely involved the full participation of key stakeholders in the resource, most notably the rural communities that usually rely on the natural resources in question for the fulfilment of subsistence needs (see, for example, Fisher 1995). In Tanga Region, the larger fragments of coastal forest[4] are usually under some form of Reserve status,[5] though many of these, particularly the mangroves, are being utilised by local communities for subsistence and by local and non-local commercial users. A strategic planning workshop with senior regional government staff identified three key issues related to coastal forest management in Tanga Region. First, the ineffectiveness of a protection and policing role for forestry field staff, who are not in sufficient numbers to control numerous, often large, scattered and remote resources. Second, the complete lack of resources for habitat restoration or development. Finally, recognition that the forest estate is under increasing pressure from various commercial pressures. The regional government staff decided that solutions needed to be explored based on an understanding of the status of the coastal forests, and the capacity of villagers to form a partnership with government to manage the forest estate jointly through collaborative management. Kipumbwi was proposed by the programme as an appropriate place to explore the process of creating collaborative forest management arrangements, as the villagers had expressed interest in working with the programme in mangrove management and as it was already within the programme as a pilot village for reef and fisheries management.

4.2 The case study context

Kipumbwi is a major fishing village in Pangani District. It consists of four sub-villages with a total of about 130 households. Adjacent villages are Kwakibuyu to the west, Sange to the south and Stahabu to the north. The population of Kipumbwi is 981 (Gorman *et al.* 1996). A large mangrove forest lies along the coast and estuary of the Msangazi River (see Map 4.1). This forest was the focus of our attention for collaborative forest management.

The investigation was conducted by a multidisciplinary team of government staff. Work had been in progress to establish village action plans for improved reef and fisheries management, for pest animal control and environmental health, for six months prior to the start of the forestry investigation work. This allowed the survey team to take advantage of a build-up of rapport and trust within the village, but also meant that some work needed to be undertaken to assess the status of institutional development (both indigenous and sponsored), before further progress could be made in collaborative forest management.

During a focus group[6] meeting with 12 women, the history of forest resource use and management was outlined. They explained that before the (Tanga) programme came there was no experience of the management of mangroves. Before villagisation and the creation of village government (in the late 1960s), villagers felt free to cut the mangroves outside the traditional sanctuary sites to satisfy their needs for boat and house building and fuelwood. There was little pressure on the forest in those times. They did not need to seek permission from the village headman (*Jumbe*) or elders. In recent years they have seen the forest under increasing pressure from traders (significantly by boat from Zanzibar, cutting poles for commercial sale), but they have felt powerless to intervene in all except their traditional sanctuary sites. In the last few years the Mangrove Management Project (another development project supporting mangrove management) has imposed a complete protection regime on the mangrove forest,[7] so that they have to seek permission from village, divisional and district government officers in order to cut anything.

4.3 Organisations for natural resource management at Kipumbwi

The villagers reported no organisations for management of natural resources, other than protection systems at traditional sanctuaries. The programme encouraged other organisations to form in late 1995, following the concern expressed by villagers over the decline in their resource base, particularly of marine resources. Two international projects are involved in natural resource management at Kipumbwi, The Tanga Coastal Zone Conservation and Development Programme and the Mangrove Management Project (MMP).[8] Both projects have established Kipumbwi as a pilot village in order to develop a strategy for improved natural resource management through local participation. The Tanga Coastal Zone Programme has been working to support villagers in the alleviation of all their identified natural resource management problems, through a number of committees. The Mangrove Management Project have been supporting the protection and planting of mangroves. A number of areas of forest are (or used to be) protected under indigenous management systems. Two mangrove areas (Kitoipi and Kwakibibi) are traditional sanctuaries, protected for spiritual worship. Elders in Kipumbwi Mtoni have traditionally been responsible for the protection of these areas. There is a strict rule that nobody may enter the area without the consent of the three elders responsible. When people come to worship and succeed with their prayers, they leave a small offering at the site.

Following a community-led planning process with programme staff (analysing priority development issues and solutions), the villagers established a number of committees to deal with management of natural resources: (1) the Lands and Environment Committee, responsible for mangrove planting and for the development of latrines in the village; (2) the Safety and Security Committee, responsible for the enforcement of fisheries and forestry regulations – this committee works closely with the Lands and Environment Committee; (3) the Mangrove Committee, responsible for mangrove planting and the control of illegal cutting of mangroves under the Mangrove Management Project; (4) the Planning and Finance Committee, responsible for the economic development of the village, through income generating activities like mariculture (prawn and fish farming) and seaweed farming; and (5) the Agriculture and Vermin Control Committee, responsible for the eradication of wild pigs that threaten agricultural production.

4.4   Analysis of the effectiveness of current sponsored organisations

Initial work in the village examined the nature of the institutional and organisational framework for natural resource management in the village, based on the following key questions: Do these recently formed committees represent a sound organisational basis for decision making on natural resource issues? Are they equitable in terms of representation and decisions made? Is village government an appropriate institution to lead collaborative management?

Two committees in particular were examined: the Safety and Security Committee (at the time primarily active in reef and fisheries management) and the Mangrove Committee (wholly sponsored by the MMP). It was felt that this analysis would provide a sound basis from which to guide the future institutional development for collaborative management of the Kipumbwi Mangrove Reserve.

The interviews in Kipumbwi Mtoni revealed that all people interviewed had heard of the Safety and Security Committee, all knew of the proposed village byelaws that were to limit the type of fishing gears that could be used and the reefs which were to be closed. Respondents also seemed aware of the main issues in fishing, the problems of dynamite fishing and the consequences of the use of small net sizes. Many were not taking an active part in committees, but felt that they had a chance to be involved in decision-making if they wished. The committees seemed to share information and issues freely for discussion, before voting on resolutions.

The survey of the effectiveness of the Mangrove Committee revealed that the products and services provided by the mangrove forests are a vital component of the livelihood strategies of the Kipumbwi villagers. They rely on the mangroves for coastal protection from erosion, building materials (ribs for boats and poles for house construction) and fuelwood. They are also a potential component of the household income of villagers, through the sale of products. Few respondents, however, had heard of the Mangrove Committee and fewer still understood its purpose. The purpose appeared to be the protection of the resource for the government. These points were further discussed at a large meeting of villagers at Kipumbwi Mjimpia. The villagers took over the meeting completely about halfway through the discussion and made a number of resolutions. First, that the Mangrove Committee should be absorbed into the working responsibility of the Lands and Environment Committee and that this committee should also coordinate all the natural resource management activities of the village. Villagers reported in household interviews that there was good communication between the committee groups, so this coordination of natural resource management activities under one committee would seem a logical and sensible development.

Second, that not enough people were involving themselves actively in the process. The women in particular were trying very hard to involve all users in decision-making, but many were simply not aware enough of the importance of the issues – more consistent and full involvement of village resource users would take more time. The recent number of significant organisational changes (changing committee responsibilities and reducing the numbers) may indicate that the villagers are learning rapidly as they proceed with decision-making, planning and action on environmental issues.

In terms of authority and accountability within a future mangrove management group (represented by the Lands and Environment Committee), they agreed that the committee would be autonomous in decision-making but remain accountable to village government. A concern not specifically addressed at this meeting, but raised during informal interviews with selected elders, was that the elders appeared to have no involvement in the process and were not involved in any decision-making capacity. This is of particular concern regarding the elders with responsibilities for management of sanctuary sites. The structures of power and authority are rapidly changing within the village with the recent democratic movement in Tanzania (there are no longer ruling party representatives alone in village government). There has also been an erosion of the power of traditional elders through the socialist movement, by centralising power and decision-making and encouraging the breakdown of tribal identity.

4.5 Establishing the full constituency of use rights: the introduction of Sange village into the management planning process

Informal meetings were held in the other settlements adjacent to the mangrove reserve (Kwakibuyu to the west and Sange to the south), to establish the full constituency of rights of access and use. Kwakibuyu residents explained that they obtained their subsistence needs from other forests. They used the mangroves very rarely and had no objection to management authority being given to Kipumbwi village. Sange village has a population of 914 (188 households) within three sub-villages. Sange villagers use the southern section of Mangrove Reserve for poles, pestles, fuelwood, salt production (in the mudflats south of the Msangazi). They have a strong interest in taking responsibility for management of the mangroves and if given authority would allocate certain areas for use, conservation and replanting.

Following further discussions with Kipumbwi, and Sange forest users, it was agreed that they would manage the mangrove forest jointly. Representatives of forest users from the two villages met again and formalised arrangements for collaborative management with representatives of respective ward and village governments, elders, forest users and the Lands and Environment Committee of Kipumbwi. The group agreed to select representatives for a Lands and Environment Committee at Sange (to complement the Kipumbwi committee of the same name) and for a smaller coordinating committee where a small number of representatives from Sange and Kipumbwi would coordinate the management activities between the two villages. They agreed to share rights equally as primary users for the whole forest, and also agreed to equally divide financial revenues. The division of revenue in this way would seem to avoid the need to define a revenue boundary for the forest (which is fortunate as there is a disputed administrative boundary between the two villages). The mangrove forest would now be called KiSa forest (rather than the Kipumbwi Mangrove Reserve) in recognition of this inter-village partnership.

4.6 Negotiation of a management plan for the resource

The management plan describes the silvicultural regime and also the institutional arrangements for forest management and is divided into a number of sections that describe the forest, the roles of partner organisations and protection and management arrangements. The crucial elements in the negotiated roles are that: (1) the forest users have exclusive rights to forest products made available through the implementation of the management plan; (2) forest users are accountable to village government but retain authority to make management decisions; (3) the Lands and Environment Committees represent the forest users; (4) the forest users can delegate responsibilities and authority to the Lands and Environment Committee and the Coordinating Committee, but they can change the decisions made by those committees or remove any members based on a majority vote in a meeting of a quorum of members; and (5) central government provide advice and assistance on demand.

The management plan includes a number of action plans that detail the management objectives, actions and individual responsibilities within a three-year time frame. The procedure and framework for action planning were based on the approach use by the programme for reef and fisheries management. The action plans form the basis for monitoring and evaluation by the forest users and the programme, of the progress with implementation of the KiSa Management Plan.

5. Discussion: an analysis of the KiSa situation

It appears that the Fisher model discussed earlier has considerable applicability in the Tanzanian context. The difficulty in the study area is that

existing institutional and organisational arrangements for resource management are weak or dysfunctional. There were never any systems of
management of forest resources at Kipumbwi or Sange, only traditional
systems of protection of selected areas as sacred groves. Even here, the
role of the clan elders in decision-making has weakened with the advent
of socialism in Tanzania. Power over resources has effectively transferred
to village government, through the influence of a one-party state.

In the case of a management system for KiSa forest, both an organisational superstructure and an institutional base are being externally sponsored. Clearly a great deal of *additional* care is required in this process
(both in establishing the management system and in monitoring the
effectiveness of the system once established), than if there was an extant
indigenous management institution with existing norms and behaviours
from which to base an organisation for collaborative management.

We can now return to the two questions posed at the beginning of
the study of the institutional arrangements for forest management at
Kipumbwi.

*Do these recently formed committees represent a sound organisational
basis for decision making on natural resource issues?* The results of the
household surveys and the subsequent village group meetings indicate that, although a relatively new organisation, the Lands and
Environment Committee seems to offer a sound organisational basis
for natural resource management. The Mangrove Committee would
be absorbed into the working responsibility of the Lands and
Environment Committee to complete its mandate for broader natural
resource management.

*Are the new committees equitable in terms of representation and decisions
made?* There were no problems raised during the household survey or
the other group meetings about this, other than the problems with
the Mangrove Committee (which resulted in a decision to dissolve
that committee). As with all new organisational structures, equity in
representation, decision-making and effectiveness will need to be
checked periodically, probably through an external and internal monitoring process. The Sange committee do not yet have sufficient knowledge of the concepts and process of collaborative management, so
they will need consistent support to build their capacity. The new
Coordinating Committee will need very careful checking, as a small
group of people are being chosen to represent a large group of users
over the management of a valuable resource. The risks of elites

dominating and/or politicising the process of decision making are high. The role of independent monitoring will be crucial.

The situation was clearly evolving rapidly as the villagers' understanding of the process and its objectives grew and the politico-administrative structure also responded (more slowly) to a newly emerging shift to democratic processes in government in Tanzania. The absorption of the Mangrove Committee into the Lands and Environment Committee is a sensible step in rationalising decision-making for mangrove management at Kipumbwi. The household survey revealed that the Mangrove Management Committee was not functioning effectively anyway. The division of responsibilities for mangrove management between three committees is also inefficient and likely to lead to confusion, so to give the Lands and Environment Committee the lead role, supported by an enforcement group (the Safety and Security Committee) would seem quite rational.

The erosion in power of the elites is probably a natural process that began with the *Ujaama* movement, though it may have been a reflection of the lack of broad-level participation in the management planning process – a point borne out in this study and further investigated by Gorman *et al.* (1996). The latter study confirmed that there was wide awareness of what the Lands and Environment Committee were doing, but that still more needed to be done to improve participation, particularly by marginalised groups, including women. The committees are now being encouraged to keep records so that attendance and participation in meetings can be monitored.

Is village government an appropriate institution to lead collaborative forest management? The present law and policy framework in Tanzania is quite unclear regarding the appropriate steps needed to empower local committees; and of the nature and security of tenure for local communities of forest under collaborative management. Precedents are being made for collaborative forest management through village government on non-reserve forest land, by gazetting as Local Authority Forest Reserve (e.g. in Babati District, see Wily 1995 and 1996). The district government appears, however, to retain management control. The relationship between the forest users, village and district government under this scenario remains unclear. Experience from other countries (particularly South Asia, see Hobley 1996) suggests that the creation of Local Authority Forest Reserves or Village Forest Reserves with power vested in district and village government (respectively), may not lead to sustainable and equitable forest use.

Tanzania does have a unique history, however, which will lead to a unique institutional and organisational framework for collaborative forest management. Thirty years of socialism (almost two generations) has brought about the institutionalising of village government as the decision-making authority in the village. Some 'indigenous' management systems have been founded on this.[9] Village government has generally, however, not functioned particularly well. Kipumbwi villagers complained of weak leadership, poor financial management and a lack of records. The political situation in Tanzania has changed from the days before socialism when clan elders and the village head had authority. The socialist government deliberately and successfully undermined this system. Changes are still occurring rapidly with the introduction of democratic processes. It is clear that the old systems relying on either elders or village government alone may not be appropriate and sustainable. Some blend of old and new institutions will guide the way forward, and the committees undertaking the process of collaborative management will need a lot of external support as the new institutions emerge and evolve. The balance of authority will probably lie between a collaborative partnership of village government, district government and forest users (represented through a committee). The legal framework does not support this structure, so experiences from pilot sites like Kipumbwi will need to be incorporated in the policy debate so that policy and law can enable rather then disable collaborative forest management. If policy and law are supportive, then collaborative management should embrace a wider constituency nationally.

5.1   Sustainability: criteria of institutional robustness

From an understanding of the KiSa forest management arrangements it is possible to analyse their likely sustainability from a comparison with criteria developed from a review of forest management organisations in Nepal and India (Hobley and Shah 1996), Uganda (Ingles and Inglis 1995), South Africa (Cousins 1995) and more widely (Uphoff 1992; Shepherd 1996; Ostrom 1990). The analysis is documented in Nurse (in press) and when applied to the context of this study draws out the following key points:

Criteria favouring the KiSa management arrangements are:

- there is a strong desire to maintain the commons as a common pool resource, rather than retain it as a state-controlled resource or private land;

- there is a strong sense of community within and between the villages (though there have been conflicts between the two villages of Sange and Kipumbwi which may resurface once substantial revenues accrue);
- there is a substantial need for the resource to satisfy livelihood needs;
- there are well-defined boundaries of access and use;
- there is a reasonably large resource (of approximately 1.3 ha per household), when compared to south Asian examples of community forestry;
- there is a strong desire to manage the resource sustainably (to benefit future generations).

Crucial areas of concern are:

- the user group is very large and cannot meet easily to make joint decisions. The representation through committees therefore is of concern, as power will be vested in a few individuals;
- there is no legal basis for collaborative management in Tanzania which results in high transaction costs and high risk for the participants in the process (particularly for the rural poor);
- there are no nested enterprises, or support networks outside government, to provide impartial advice to the forest users (as for example, in India).

Clearly, external support will be required for a number of years, particularly to monitor the effectiveness of sponsored institutions, and to transfer the lessons learnt (successes and failures) to government for the policy reform debate.

5.2 Monitoring the effectiveness of these new organisations and institutions

Jackson and Bond (1997) proposes (quoting from Fisher) that monitoring of socio-economic criteria of collaborative management should focus on three categories of concern: well-being (quality of life and economic factors which provide access to material goods); equity (how well-being is distributed fairly to different individuals and groups); and risk. There is a crucial need to monitor the effectiveness of the KiSa committees in their decision-making, representation and ability to resolve conflict without external support. The monitoring will be undertaken as a part of the management process and by all partners in the management plan. A monitoring plan is now being developed by the programme to satisfy their needs. The users are monitoring a number of key factors using records that have been introduced and supported by the programme. These records will allow the forest users and outsiders to monitor progress and allow the users to be accountable to the other partners as expressed in the management plan.

Institutional theory suggests that organisations are created to take advantage of opportunities offered by institutional arrangements (in this case, in the exploitation of a natural resource) (North 1990). The KiSa forest management committees represent a delicate and dynamic balance of cognitive, normative and regulative elements, that are essentially externally sponsored. Whether these elements are in the appropriate admixture to provide the necessary stability and meaning to this new institutional balance based on coercion (through sanctions), incentives (direct project support and potential improved rights of access) and participation (with the state, the project and each other) remains to be seen. A particular challenge is whether the emerging new role of the state (as providing advisers and extension services rather than protection and enforcement agents) can help in the long term to overcome the high transaction costs of the KiSa group, of time spent by villagers in meetings and in enforcement patrols. The progress of the new organisations and institutions will be watched with interest.

6. Conclusions

Tanzania has a unique set of conservation challenges that are a reflection of a unique historical, cultural and political background. The information gathered during the development of a methodology for coastal forest assessment, and subsequently during more detailed investigation at the Kipumbwi pilot site, lead to the conclusion that there is great potential in Tanzania for an approach to resource conservation and management based on collaborative forest management. The current circumstances in Tanzania are in common with many other developing countries: there is a decreasing natural resource base; there are many forest sites of international importance for conservation that are under severe threat from commercial users; the rural communities depend heavily on forest resources to fulfil their subsistence needs; and government does not have the capacity to protect and manage all forest resources through an expansion of the reserve network (even with donor assistance), unless management responsibilities are shared with local communities.

The organisational and institutional structure for forest conservation under this scenario will be uniquely Tanzanian (and unique to each

site in Tanzania) and needs to be responsive to a rapidly changing political, economic and cultural environment. This challenge lies ahead, but
several lessons can be drawn from experiences in other countries, particularly south Asia, where experience of participatory forestry has been
gained over the last 20 years. Scheinman and Mabrook (1996) suggest a
mixture of old and new elements – perhaps old traditional, recent
socialist and new democratic – will form the appropriate new management institutions. Because of the new nature of these institutions, careful support will be required from external agencies to ensure equity,
self-reliance and sustainability in institutions, and sustainability in
conservation and management of the resource.

Notes

1. The following individuals supported the authors in providing continued
   innovation and commitment in exploring an approach to collaborative
   management: Lucian L.J. Massawe, Regional Coordinator, Mangroves,
   Coastal Forests and Wildlife; Ally Mwemnjudi, District Mangroves Officer
   and Catchment Forestry Officer, Pangani District; Agnes Mfuko, Community
   Development Officer, Pangani District; Michael Mfuko, Community
   Development Officer, Muheza District. Without the support of these team
   members none of this work would have been possible.
       The authors also wish to thank Chris Horrill (Tanga Coastal Zone
   Conservation and Development Programme), Ed Barrow (Senior Programme
   Adviser, IUCN, Nairobi) and Mike Harrison (Senior Consultant, LTS
   International, UK) for valuable comments on earlier drafts. This work was
   undertaken while Mike Nurse was under contract to IUCN and responsible to
   IUCN and the Tanga Authorities. The comments represent the views of the
   authors and not necessarily those of IUCN, the Government of Tanzania or
   of the donor, Irish Aid.
2. For an explanation of the concept and methodology of action research as
   applied in collaborative management, see Jackson (1993).
3. Mike Nurse provided periodic technical inputs to the forestry component of
   the programme from June 1996 to January 1998, under contract to IUCN.
   These periodic inputs are part of wider technical assistance support under
   the Tanga Coastal Zone Conservation and Development Programme.
4. Coastal forests are those on the Tanga coastal plain between the montane
   forests and the ocean. They include sub-montane forest, coastal thicket and
   mangroves below an altitude of 700 m (adapted from Burgess and Muir
   1994).
5. All mangroves are gazetted and under the authority of central government as
   Reserves. Utilisation for small-scale use can take place under permits issued
   by the District Mangroves Officer. An agreement for collaborative management has no precedent and requires the approval Village, District and
   Regional Authorities and the Director of Forestry.

6. Focus groups are groups within the community who share common interests or concerns for resource management, e.g. women, rich farmers, fuelwood collectors.
7. In fact, the District Mangroves Officer (who now represents MMP in the village) imposed a protection regime on his arrival in the area in 1986. The forest was severely degraded at this time (*sic*).
8. The Mangrove Management Project (MMP) is a component of the larger FINNIDA/Government of Tanzania Catchment Forestry Project. MMP has a remit for all mangrove forests in Tanzania.
9. A rapid assessment of a selection of forests and villages in Tanga Region revealed a number of indigenous management systems, some of which were developed and controlled by village government (Nurse 1996).

5

Organisations for Participatory Common Property Resource Management

Madhusree Sekhar

1. Introduction

Participatory natural resource management has become the new catch-phrase in tackling environmental problems, particularly those arising from the use of common property resources (CPRs). Being limited in supply but accessible for community usage, their appropriation by one user affects that of others. The resulting externalities associated with individual actions often give rise to environmental problems, threatening resource sustainability. Keeping in view the interdependencies created by their nature, there is, therefore, a need for participatory management of CPRs.

The emphasis on participatory approaches arises from the wisdom that local communities not only understand their problems best, but also have the solutions. This makes them able participants as beneficiaries and also as providers. Within the natural resource sector, the participatory approach received an impetus following the Rio Earth Summit where it was accepted as an integral part of the sustainable development process (Dearden *et al.* 1999). An important dimension of participatory resource management is the manner in which the rules governing resource utilisation are formulated and the way the process is organised. Local organisations are an important mechanism for generating and focusing participatory management of community resources like CPRs. Recognising this, one current thrust in the development debate is the role of local organisations (both community-based grassroots initiatives and externally induced grassroot-focused organisations) in the development process (Esman and Uphoff 1984; Korten 1986; Attwood and Baviskar 1988).

Participatory resource management thus involves a process of organised management. Implicit in this is the manner in which institutional conditions shape and determine users' capacity as resource managers. Studies on local organisation have generally addressed certain normative conditions like empowerment, participation and equity, or have tried to view it as a way to improve institutional efficiency for the purpose of service delivery (Uphoff 1988). But do different organisational initiatives at the grassroots exhibit different participatory characteristics in a community? What can be learned, if anything, from the different initiatives? What lessons can be drawn in terms of policy interventions?

Participatory resource management is justified by the need to ameliorate degradation of our natural resources. However, the tendency of humans to overuse common property or public goods implies that socially optimal outcomes in managing resources accessible for community usage may be undermined, because of incentives for 'free-riding' (Baumol and Oates 1988). To prevent this, there is a need for institutional interventions that provide incentives for participatory management, while checking over-exploitation of the resource. Individual incentives could be in terms of usufruct rights, while incentives for communities could arise due to improved condition of resources, improved tree coverage, reduced soil erosion and increased soil fertility for agricultural activities. Organised participatory resource management thus does not merely have an environmental impact (protection and preservation of the resource), but in the process adds value to the resource.

The idea of organised participation follows from seminal work on local organisations by Esman and Uphoff (1984). Their emphasis on a bottom-up approach to development suggests a major role for the local population in influencing decisions that affect them and in receiving a proportion of any benefits that might accrue. In practice, this involves sharing tasks with others in the same group or with other interest groups at micro and macro levels within the framework of people's (local) organisation. There are definitional ambiguities about what constitutes a local organisation, but they are commonly referred to as village-based units which may be an outcome of the locals' own initiatives or any externally induced initiative either from the government, including local government institutions, or from any non-government body (Uphoff 1982). Taking this interpretation as the basis, the purpose of this chapter is to analyse the process of participatory resource management as mediated through local organisations.

Significant attention has been given to the goals and activities of local organisations in the process of participatory development (Ghai and Vivian 1992). The issue has been viewed in a wide variety of contexts to span a broad range of organised participatory options (Dearden *et al.* 1999). Broadly, three types of organised participatory relationships can be identified. They are:

1. *Regulatory*: formalised relationship externally induced and guided by state/government interventions to meet predetermined objectives – compliant cooperation lacking in people's initiative (people participate to implement organisation activities/programmes).
2. *Conjunctive*: participatory partnerships in a community arising from the locals' assessment of problem(s) resulting in plan formulation and implementation – cooperation conditional to each member conforming to the organisation's regulations guiding individual behaviour.
3. *Voluntary*: self-help informal initiatives of people arising out of their own needs and a recognition of their dependency on each other – people's developmental relationships independent of external inducement.

The three organised participatory relationships thus range from formalised patterns where the linkages are strong (regulatory type) to more voluntaristic cooperation where the participatory process is guided by more informal sets of norms and practices. However, the question about the type of micro-level organisational arrangement needed to make the participatory process effective still remains open. While seeking to analyse this issue, the chapter argues that local organisations, being location-specific and having a membership drawn from among the people, could enlist the locals' support. But since different organisational situations have different impacts on people's cooperation, it is vital to comprehend the differences so that the best solution could be reached.

Insights are drawn from the Joint Forest Management (JFM) efforts being carried out in the State of Orissa (India) as part of its social forestry programme. JFM represents an historical shift towards decentralisation of forest management in India through the New Forest Policy of 1988. (See Yadama and DeWeese Boyd, and Jeffery *et al.*, chapters 6 and 9 this volume, for more details on India's JFM programme.) The discussion that follows is in three sections. Section 2 presents some methodological issues; section 3 analyses the process of organised participatory management of CPRs by studying different types of local organisations involved; and section 4 outlines a path for more effective involvement of people in managing community resources.

2. Methodological issues

This chapter represents a qualitative and quantitative analysis drawing insights from a field study in three selected villages in Orissa (India). Village selection was purposive, taking into consideration the type of local organisation operating therein. In Village A, a non-governmental people's organisation (NGPO) operated – the Friends of Trees Organisation. In Village B, a government engineered (induced) people's organisation (GEPO) was active – the Village Forest Committee. In Village C a traditional management regime (TMR) existed – a village council. This categorisation was on the basis of the 'initiative' for their constitution – the NGPO was an outcome of an individual or group effort originating in the village itself; the GEPO was created through government initiative; and the TMR represented organisational conditions that had evolved over a period of time without any specific interventions from any group, person or government. These three types broadly represent the typology of organised participation suggested earlier in this chapter: the GEPO represented a regulatory system, the NGPO a conjunctive system and the TMR an informal voluntary system.

The three study villages were located in the forest-scarce region of Orissa (the coastal district of Puri) and, as such, while having largely similar socio-economic conditions, the locals' dependence on the limited forest resources available in their vicinity was quite high. All three were small villages ranging from 65 to 108 households, consisting mostly of small farmers. All three had a community component of the social forestry afforestation programme over which the villagers had rights of usage and which called for their collective action for its protection and preservation. A comparative profile of the CPR management situations in the three villages is shown in Table 5.1.

This research project attempted to explore how different structural conditions affected resource management, by taking the type of organisation as an independent variable. Subsequently, the outcomes of management efforts were examined from the users' perspective in terms of their participation in the process; the continuation of the management strategies in the three sample situations (sustainability of management practices); and, the sharing of direct produce and other indirect benefits from the CPRs on an equitable basis.

It needs to be stressed that the analysis in this chapter is limited because it concentrates on just three villages. Time and financial constraints precluded the possibility of a more extensive field survey. Furthermore, data collection was a one-time exercise and there has not been an opportunity to visit the field area again since 1992, when

Table 5.1 Profile of study villages

	Kesharpur (Village A)	Tikatala (Village B)	Lodhachua (Village C)
1. Demographic features			
a. Size of user community	Small (65 occupied households)	Small (83 occupied households)	Small (108 occupied households)
b. Population composition	Heterogeneous (dominated by *Khandayats* and other backward classes)	Heterogeneous (dominated by *Khandayats* and other backward classes)	Heterogeneous (dominated by *Khandayats* and other backward classes)
c. Population density	Moderate	Moderate	High
d. Economic composition	Mostly small farmers and landless households	Mostly small farmers and landless households	Mostly small farmers and landless households
2. Empirical data collection	Census survey (all occupied households) Structured questionnaires administered to any adult in the household	Census survey (all occupied households) Structured questionnaires administered to any adult in the household	Census survey (all occupied households) Structured questionnaires administered to any adult in the household
3. CPR Attributes			
a. Type	Community component of Social Forestry (regeneration of degraded forest)	Community component of Social Forestry (village woodlot)	Community component of Social Forestry (village woodlot)
b. Legal status	Protected forest	Village forest	Village forest
c. Scarcity (forest)	High	High	High
d. Distance of village from nearest significant forest (approx.)	5 km	10 km	5 km

the field survey was carried out. Recognising these limitations, it is emphasised that the purpose here is not to make any generalisations about local organisations in the process of participatory resource management. The objective rather is to provide broad organisational guidelines for development initiatives requiring community involvement.

3. Resource management: some empirical findings

3.1   Representation and decision-making

It was observed that the success or failure of organised participatory strategies for governing community resources depended to a large extent on the entities that were vested with responsibility for 'rule formulation' and 'rule enforcement'. This was usually the executive membership or the leadership of the organisation. The prospects for securing local coop-eration and rule confirmation for managing and preserving the resource improve if there is wider representation of the community in the organ-isation. In this regard, the local NGPO was relatively less representative and less democratically constituted than the other two types. It was pri-marily the village elite (either the socially and economically better-off or the educated) who were members of its executive body and were the regulatory decision-makers. They formed the village's 'brain machine'. They were usually the first to understand the impact of forest degrada-tion and the need for its preservation with the community's help and so provided the leadership for forming the NGPO. But by virtue of their initiatives in the organisation's formation, they retained the loyalties of the membership, and the choice of leadership in the NGPO was largely arbitrary. The leadership pattern in the other two types of organised strategies was more representative as it was the outcome of people's choice. In the case of the GEPO, the choice was made through an infor-mal election process, with each hamlet of the village choosing its own representative. Under the TMR (traditional management regime), the villagers informally gave their unanimous support to some village elders (respected locals) who looked after village affairs (see Table 5.2).

It was also noted that the decision-making structure within the organisation was a key precondition for the stability and success of the participatory process. The user-community was likely not to cooperate if they were not consulted by the organisation's leadership and not informed about its decisions. Although the NGPO's rules provided that decisions taken by executive body had to be approved by the general body of members from the user community, this did not happen in practice. Villagers complained that the organisation had stopped call-ing regular village meetings and that the participatory decision-making process had broken down. The organisation mainly operated through its executive body and a few other prominent individuals in the village.[1] On the other hand, although the GEPO also operated through its committee (executive body) members, they were the chosen repre-sentatives of the people.[2] Of the three cases studied here, the third type

Table 5.2 Comparison of CPR management systems in study villages

CPR management	Kesharpur (Village A)	Tikatala (Village B)	Lodhachua (Village C)
a. Reasons for management	Increasing soil erosion; decline in soil fertility, severe fuelwood and small timber scarcity; environmental concerns	Severe scarcity of fuelwood and timber	Acute shortage of fuelwood and small timber; environmental concerns
b. Type of village organisation	Non-government people's organisation (NGPO) Friends of Trees and Living Beings	Government engineered people's organisation (GEPO) – Village Forest Committee (VFC)	Traditional management regimes (TMR) Informal Village Council
c. Legal status of organisation	Registered body	Satutory	Informal
d. Initiative for organisation's formation	Group or local individuals (leaders)	Social Forestry Directorate (Government)	No formal initiative
e. Decision-making structure	Executive (elected representatives)	Committee plus assembly	Committee
f. Protection system	Initially community vigilance, later voluntary active patrols	Paid watcher and community vigilance (no active protection)	Voluntary protection through stick rotation (Thengapalli)
g. Perceived benefits	Comparatively few direct benefits (due to restrictive access rules imposed by the organisation) mostly indirect benefits like regular monsoons, increase in soil fertility and stability in streamflow	Mainly fuelwood and timber	Mainly fuelwood and timber
h. Mobilisation of users	High (visible through their environment consciousness)	Low	Moderate

(TMR) provided the best illustration of a participatory decision-making structure. It primarily followed a consensus-based approach wherein both the committee (executive leadership) and assembly or *Mouja* (one representative from each household) actively participated in planning resource use and management. The executive body prepared action plans, but its decisions were adopted in consultation with the assembly.

A related issue is the manner in which the rules of operation are interpreted and enforced by the organisation. These rules formalise people's activities and their response to the resource management efforts. The leadership in the NGPO was over-zealous in enforcing the rules for forest preservation and maintenance. This was reflected in the management conditions provided by the organisation. For instance, during the initial years of its activities, the NGPO leadership decided to enforce a strict prohibition on the locals from using the forest with a view to promoting its regeneration. Entry into the protected area by anyone carrying an axe or sickle was banned. Violation of this rule could result in the person being fined. Grazing of animals was also prohibited initially to allow for natural regeneration. Subsequently, restricted grazing was permitted on the foothills only. While these measures may have been appropriate from the conservation point of view, with the organisation receiving public recognition (it received the Priyadarshini Award from the Indian government) they were not responsive to local needs and gave rise to discontentment. The TMR's actions in Village C, being more consensus-based, were more user-friendly. Although there were strict regulations to check free-riding, members of the user-community enjoyed certain recognised rights with regard to resource appropriation. They had free access to collect dried leaves and twigs for fuel. Regular coppicing of the forest was carried out and the produce thus collected was equally distributed. In case of disagreement, remedial measures could be taken at village meetings. The GEPO's activities also had the support of the people, but this was more passive. The community members did not have any direct role in managing the resource. This was the responsibility of the committee members (people's representatives). The community was generally informed about what was happening or about to happen and consulted if necessary. In accordance with government specifications, the community only had usufruct rights. Protection of resources was carried out formally through paid labour and informally by way of social fencing (villagers kept an informal watch to ensure no outsiders entered the woodlot).

3.2 Accountability

The study showed that if the leadership was accountable to the user-community, it was able to evolve more user-friendly management conditions and consequently, generate public support for the organisation. Accountability was assessed using two criteria: (1) by examining the selection process of the organisation's leaders; and (2) by evaluating the organisation's decision-making structure. Though the TMR leadership was made up of village elders (hence no selection was involved), accountability was ensured through its decision-making structure. The practice of regular general body (*Mouja*) meetings provided an opportunity to keep the community informed and for the community to review the leaders' role.[3] On the other hand, accountability of the GEPO's members was ensured through their selection by the user-group. Committee membership was for a term of five years, after which they had to be re-selected by the community; in case of public dissatisfaction, they could be removed earlier. In this case, however, village meetings with the leaders were not held on a regular basis, and day-to-day accountability was absent. The situation was quite different in Village A. The arbitrary nature of NGPO leadership and the absence of regular meetings with the villagers resulted in low user-involvement in decision making and their growing suspicions about the organisation's objectives.

3.3 Participation

Participation was analysed in three types of resource management activities: rule formulation (planning), rule enforcement (implementation) and resource maintenance. Seven indicators were identified, of which five were direct participation variables: (1) sensitisation about the need for afforestation, (2) involvement in tree planting activities, (3) involvement in resource maintenance, (4) involvement in protection, and (5) degree of financial contribution. These broadly correlated with the action phase of CPR management. The two other indicators considered were 'nature' and 'levels' of involvement in the afforestation activities. These were variables that indicated the *type* of organised participation existing in the study areas. 'Nature of involvement' defined the manner of user-participation in resource management – whether direct participation or indirect, and, if indirect, whether through elected representatives or local leaders. 'Level of involvement' was used to determine the extent of people's involvement – whether voluntary or induced (through material incentives); and if voluntary, whether involvement extended to formulation of policy or was restricted to implementation of decisions.

The study revealed that in Village C, the majority were engaged in managing the adjacent forest that was used as the village CPR. They were involved in rule formulation (policy-making) and its implementation (enforcing regulations). Variations in terms of caste and class had no visible impact on the community's involvement. Collective decisions concerning the resource management were made in village meetings (*Mouja*), which were attended by at least one member from every household. In Villages A and B, on the other hand, people's participation in rule formulation was not very evident. This was mainly the responsibility of a small core group – the elite in Village A and the chosen representatives of the people in Village B. People's involvement was mostly confined to rule implementation and resource maintenance.

3.4 Sustainability

Sustainability, in this analysis, is viewed as the continued and effective administration of the resource, leading to asset creation (forests) that benefited all stakeholders. It thus implies the sustainability of the physical resource as well as sustainability of the collective management system. However, given that resource sustainability depends on continuity of the resource appropriation pattern determined by the system (Kant *et al.* 1991), sustainability of the collective management practices assumes primary significance in any endeavour to protect and preserve a CPR. The issue was analysed first by examining the capacity of user groups to influence decisions affecting resource management, in order to see whether management efforts were responsive to changing demands; and, second, by studying local institutional capacity to support and sustain the user-group's cooperation for maintaining the resource. The first variable was operationalised by studying the people's role in decision-making for managing the resource at various stages – during the initiation of the programme; the selection/identification of the site for afforestation activities and the type of species to be planted; the formulation of regulations guiding the maintenance of the resource and its utilisation; and the selection of leaders/executive members of the local organisations responsible for management. The second issue was examined by considering the organisation's autonomy and the accountability of its leadership to the user-community.

As far as the capacity of the user-group is concerned, the analysis revealed that the best results were in Village C where the consensus-based approach of the TMR allowed for greater user-involvement in

Table 5.3 Perceived user involvement in decision-making in sample villages

Involvement in decision-making	Village A		Village B		Village C	
	Whether consulted	Views respected by organisation	Whether consulted	Views respected by organisation	Whether consulted	Views respected by organisation
Not at all	8 (12.3)	11 (16.9)	—	—	1 (0.9)	1 (0.9)
Rarely	26 (40.0)	35 (53.8)	4 (4.8)	1 (1.2)	3 (2.8)	3 (2.8)
Occasionally	27 (41.0)	15 (23.1)	47 (56.6)	62 (74.7)	12 (11.1)	21 (19.4)
Mostly	1 (1.5)	2 (3.1)	28 (33.7)	20 (24.1)	59 (54.6)	83 (76.9)
Always	3 (4.6)	3 (3.1)	4 (4.8)		33 (30.6)	
Total	65 (100.0)	65 (100.0)	83 (100.0)	83 (100.0)	108 (100.0)	108 (100.0)

Note: Figures in the parentheses indicate percentages and outside the parentheses indicate the number of households.

decision-making (Table 5.3). Nearly 85 per cent of the households said that they were consulted. But in Villages A and B, the organisational arrangements allowed for limited user-involvement in decision-making. In Village A nearly 52 per cent felt that they were rarely consulted. In Village B, though there was low user-involvement in decision-making, the situation was slightly better as the community was consulted in the event of urgent/important matters which required their consent before an action could be taken. Such occasions arose mostly following theft from the protected forest, when the nature of punishment was to be decided, or while selecting the GEPO's committee members. The low community involvement in decision-making in Village A was reflected in its gradual disenchantment with the management strategy initiated there. On the other hand, because of better user-involvement in the resource management process in Village C, their commitment to the organised efforts to maintain and preserve the resource was higher, thus contributing to the sustainability of the participatory system.

Studies have shown that local organisations are in danger of losing their autonomy, either to the government or to the local elite/leaders (Esman and Uphoff 1984). The purpose here was to examine the extent to which these local organisations minimised external influences, so that instead of being detrimental, the checks on their autonomy could be productive for their performance. The GEPO operating in Village B was a government initiative. It was undoubtedly an important institutional innovation to build the locals' capacity in collective management of their woodlots. But autonomy was a casualty, and the organisation, in practice, followed government guidelines. Besides, the Village Forest Worker, a government employee and an ex-officio member of the executive committee, was primarily responsible for supervising this local participatory strategy. Consequently, the organisation lacked initiative, and effective self-help endeavours were absent. This reflected a passive organisational set-up where the community had   only functional responsibilities exercised in accordance with government norms. On the other hand, in Village C, public involvement was ensured through the general assembly of the TMR (*Mouja*), resulting in greater local assertiveness in resource maintenance and utilisation. This in turn fostered more autonomy in the organisation's performance. The NGPO in Village A also enjoyed autonomy from government control, reflected in the initiatives taken by it to generate forest-consciousness among the locals to mobilise them to adopt forest preservation activities and to take up the protection of degraded forests in their locality. But, as mentioned earlier, the absence of a role for the community in decision-making had a debilitating effect on the participatory strategy. This indicates that autonomy from external control without countervailing pressures due to internal community monitoring can negatively affect an organised participatory strategy.

In terms of accountability of the leadership to the user-community a similar trend was also observed. As discussed earlier, the leadership under the TMR system had the unanimous support of the community, as was also the case in Village B. But the situation was quite different in Village A, where a few individuals controlled the management process and the people, in general, were critical of its decisions. This appeared to have a negative affect on the sustainability of the management system in the village.

3.5 Benefit sharing

Given the significance of forests in rural lives, popular involvement in resource maintenance can be sustained only if the locals have recognised rights over the resource and expect to derive benefits equally from them. Accordingly, the issue was analysed here from two perspectives: access of the community to the resource, and the types of benefits enjoyed by them. The former was examined in terms of the method employed for biomass extraction. The latter was assessed by considering six broad types of benefits: (1) increased employment; (2) increased microeconomic activities through collection and sale of forest produce; (3) availability of fuelwood/fodder/food for household consumption; (4) time saved in collecting fuelwood; (5) changes in soil fertility (availability of green manure and reduction in soil erosion); and (6) changes in climatic conditions.

Regarding access to the produce, it was observed that in all three cases there was no discrimination, and everyone had equal usufruct rights. But when it came to exercising the right, variations were observed. In Village A, following the successful mobilisation efforts of the NGPO, the locals wholeheartedly supported the afforestation activities and even sold their small ruminants (goats and sheep) to protect the forest and facilitate its natural regeneration. The opportunity cost of protection for the people, particularly the poor, was quite high. But, because of extensive restrictions laid down by the NGPO, they were getting low returns as most of them refrained from extracting any produce from the protected patch. In Village B, on the other hand, the villagers were allowed to collect the fallen leaves and twigs from the forest floor on a 'sweep and carry basis' by headloads, a practice mainly followed by the poor (the landless and small farmers). A similar trend was observed in Village C. Besides, in this village, annual coppicing of different patches of the

Table 5.4 Perceived benefits from CPRs in the sample villages

Types of benefits	Village A (N=65)			Village B (N=83)			Village C (N=108)		
	Unsatisfied	Somewhat satisfied	Satisfied	Unsatisfied	Somewhat satisfied	Satisfied	Unsatisfied	Somewhat satisfied	Satisfied
1. Increase in employment opportunities	62 (95.4)	3 (4.6)	—	7(8.4)	16 (19.3)	60 (72.3)	8 (7.4)	17 (15.7)	83 (76.8)
2. Increase in micro-economic activities	49 (75.4)	5 (7.6)	11 (16.9)	76 (91.6)	—	7 (8.4)	96 (88.9)	12 (11.1)	—
3. Increase in fuel/fodder availability	17 (26.1)	38 (58.5)	10 (15.4)	7 (8.4)	17 (20.5)	59 (71.1)	5 (4.6)	31 (28.7)	72 (66.7)
4. Time saving in collecting fuelwood	7 (10.8)	33 (5.08)	25 (38.5)	7 (8.4)	5 (6.0)	71 (80.5)	11 (10.2)	14 (12.9)	83 (76.9)
5. Increase in soil fertility	—	8 (12.3)	57 (87.7)	2 (2.4)	16 (19.3)	65 (78.3)	6 (5.6)	13 (12.0)	89 (82.4)
6. Positive changes in climatic condition	—	—	65 (100.0)	9 (10.8)	15 (18.1)	59 (71.1)	—	9 (8.3)	99 (91.7)

Note: Figures in the parentheses indicate percentages and outside the parentheses indicate the number of households.

woodlot was organised and the produce thus collected equally shared by participating households (each household was expected to participate in this exercise). If the economically better-off did not take their share, it was sold at a very nominal rate and the proceeds were remitted to the village fund to be used for community activities.

The nature of benefits accruing to the community varied in the three villages. In Villages B and C, where the locals exercised their usufruct rights and collected produce from the protected forest, more direct benefits were perceived (Table 5.4). However, perceptions in Village A of indirect benefits arising from improved environmental conditions such as increase in soil fertility, rising water-table, etc. were quite high. Some economic gains were perceived in Villages B and C, where existing organisational arrangements allowed for paid labour in planting activities. However, with regard to the use of non-timber forest produce as an income-generating activity, not much benefit was perceived in the three villages. This can mainly be attributed to two factors: first, due to acute fuel scarcity in these areas, the trees planted as part of afforestation activities were primarily fuelwood producing varieties; and second, the three villages had predominantly non-tribal populations who were less dependent on forest products as a livelihood source.

4. Conclusion: structuring organised participation

Depending on the process by which rules are made and interests internalised, different local organisations create constraints, provide opportunities and confer legitimacy differently. While the NGPO appeared to be more effective in mobilising the community and taking initiatives for promoting resource preservation, it seemed unable to secure sustainability of the collective strategies. On the other hand, the TMR enjoyed better community support, but it would be naïve to rule out the possibility of the resource being abused if the collective (consensus-based) decision-making process were to break down. The GEPO, specifically constituted by the government to manage forests used as community resources, did provide the necessary legitimacy. It capitalised on the advantages of decentralised governance while, at the same time, being a membership organisation. It was able to ensure the government's accountability to local problems as well as the leadership's accountability to the people. But being the outcome of government efforts and allowing only representative participation, this type of local organisation was seen to lack autonomy and initiative. It follows that no single type of organisation would optimise the goals of community resource management. There is a need for an integrated perspective, implying a collaboration of all concerned, not only within the user-community but also between the community and the government.

In studying the range of organised participatory management options, it must be recognised that this analysis is only a first attempt to determine appropriate structures for participatory resource management. Changes in characteristics of the user-community and the resources may have a major impact on the organised participatory process, which is beyond the scope of the present study. However, the study has highlighted various issues which may receive priority in future planning for participatory resource management in India, and possibly also in other poor developing countries faced with similar conditions of resource scarcity and a perceived need for user involvement in their management. The main issues that have emerged from this analysis are that local organisations have to be rule-based rather than personality-based; that they should be representative of the user-group, should encode different power-relations and be accountable to the community; and, that they should have greater adaptability to local challenges and needs.

Notes

1. The memorandum of association of the Friends of Trees Organisation provides that its membership will be open to all individuals interested in environment preservation on payment of an entrance fee of 1 rupee. Although based at Kesharpur, this NGPO has a membership drawn from 22 villages representing *five gram panchayats* (eight hill forest villages which were directly involved in its protection and 14 others who indirectly contributed towards its protection). Most of the villagers from these villages are its members. They constitute the general body whose main responsibility is to approve the activities of the organisation. The executive committee of the organisation consists of a president, four vice-presidents, a secretary, a treasurer and 18 members representing the different villages. Since the present study focused at Kesharpur, only the executive members from this village (numbering 6) have been examined.

2. As per the guidelines laid down in the Orissa Village Forest Rules, 1985, the Village Forest Committee (the GEPO) must comprise the *sarpanch* of the *gram panchayat* under which the village comes, the ward members and such other persons selected by the villagers. Following the recommendations made in the Mid-Term Review of the State's Social Forestry Project in 1991, the rules regarding the chairmanship of the VFC have been relaxed. Very often now the VFC chairman is a person from the village itself. If the *sarpanch* is from the village where the VFC is formed, then he is made the chairman. The rules stipulate that the members chosen by the villagers shall

be not less than three and not more than five. However, there is some ambiguity in this regard. In practice, it is observed that the VFCs generally consist of 8 to 10 members in total. These are the non-official members of the organisation. In addition, the VFC has two officials as its ex-officio members – the concerned Village Forest Worker and the Revenue Inspector.

3. The meetings with the general body were held at least once in a month, or as and when necessary. It was binding on every general body member of the village council to attend *Mouja* meetings. If a person could not attend, he had to inform the concerned ward member or had to send some one else from his household as his representative. Otherwise, he was liable to be fined.

6

Co-management of Forests in the Tribal Regions of Andhra Pradesh, India: A Study in the Making and Unmaking of Social Capital

Gautam Yadama and Margie DeWeese-Boyd

1. Introduction

The debate surrounding natural resource management in South Asia is no longer centred on the issue of whether management is to be participatory or non-participatory. That question appears settled: natural resource management is to be participatory. For almost a decade now, participatory management philosophy has been the dominant guiding force in the forestry sector in India, Nepal and Bhutan. For instance, community and private forestry are the dominant components in Nepal's Master Plan for the Forestry Sector. These two components are expected to absorb 47 per cent of all investments made in the forestry sector until 2010 (CPFD 1997). The Joint Forest Management programme (JFM) is an effort in participatory natural resource management (PNRM). JFM occupies a central place in India's efforts to conserve, manage and regenerate forests via state and community partnership arrangements. Our chapter looks beyond the basic arguments for and against co-management of forests. We will focus on the unique problems and prospects of deploying forest co-management in tribal communities in the Eastern Ghats region of Andhra Pradesh, India. In so doing, we will explore the problems and prospects of successfully involving tribal communities in JFM, and the relevance of social capital to that process.

Social capital is important for governing community forests due to its potential to reconcile individual and collective rationality, and thereby enable members of a community to overcome dilemmas of collective action. We will argue that where collective decisions and actions are called for – such as the case of jointly managed forests – they are more likely to be successfully undertaken in groups where social capital is high. Social capital has also been identified as important for co-production of public goods – goods produced in collaboration by the state and citizens (Ostrom 1996). In this context, our chapter will underscore the importance of supplying social capital at the micro-institutional level.

The chapter will unfold as follows. Our first task is to describe the extent to which the tribal household economy is exclusively dependent on resources from the forests. Forest co-management strategies take on added significance as one grasps the degree of tribal dependence on forests. Next, we will present the origins of conflict and contention between tribals, non-tribals and the government in the form of forest and revenue departments. Having provided the background, we will then outline the JFM programme with reference to Andhra Pradesh and the paradigmatic shift it constitutes in Indian forest policy. Unlike other forestry projects, the JFM programme attempts to build village-based institutions capable of self-governing community forests, and at times co-managing these forests with the forest department. Co-management is based on trust and reciprocity within and between tribal communities and the forest department. Social capital reserves are essential if such a joint venture is to be successful.

We extend the discussion of social capital and its significance to the production of JFM in three ways: (1) we will disaggregate and delineate the complex set of interrelationships that are broadly classified as social capital. This will allow for increased conceptual utility and explanatory power of social capital theory. (2) By providing examples, we will illustrate the ways in which social capital has been fostered, and how it subsequently enables tribal communities to take up co-management of community forests in this region. (3) We will discuss the ways in which social capital has been undermined in recent months and how it might jeopardise many of the established forest co-management arrangements in the Eastern Ghats region. By developing our chapter in this way, we attempt a convincing argument that while social capital is essential for co-management of forests it is extremely difficult to supply as well as sustain, and can be easily eroded.

2. Forests and tribal household economy

It is estimated that there are 66.5 million tribals in India (Maheshwari 1990). With some exceptions, the majority of tribals are forest-dwellers.

In spite of many competing demands and pressures on forests, tribals continue to depend on forests for their livelihood. Forests and forest resources, primarily minor forest products – or non-timber forest products (NTFPs) – play an important role in the viability and survival of tribal households in Andhra Pradesh and elsewhere in India. Sustainable use of forests and forest resources is a necessity for tribal households in India, because of the importance of forests in their social, cultural and economic survival (Tewari 1989). Estimates of the revenue contributions of NTFPs in India vary considerably. Some estimate that NTFPs contribute US$ 208 million to the Indian economy, while another calculation places the revenues from NTFPs at US$ 645 million (Lele *et al.* 1994). Yet another estimate proposed by Poffenberger (1990a) estimates that the total annual value of NTFPs from the central Indian tribal belt exceeds $500 million. All the estimates, while varying from one another, lead to the firm conclusion that collection and processing of NTFPs is an economically significant activity for forest-dependent tribals. In sum, tribal dependence on forests for NTFPs and fuelwood is extensive.

Recent studies conducted in the tribal regions of Bihar, West Bengal and Karnataka offer further empirical evidence for the extent of dependence of tribal households on NTFP collection (Rao and Singh 1996). For example, in a study of Soliga tribal households, Hegde *et al.* (1996) found that the income contribution from the collection of NTFPs is disproportionately greater than the time spent in collecting the products. In Andhra Pradesh, tribals collect a large variety of NTFPs.[1] One study estimated that income from the sale of NTFPs in Andhra Pradesh constitutes anywhere from 10 to 55 per cent of total household income and this dependence increases markedly in the very poor households (Roy Burman 1990). Household studies conducted in different villages of high altitude tribal regions of Vishakapatnam district further illustrate the high degree of tribal dependence on forests. Data indicate that income from forests comprised more than 50 per cent of the household income. Even in areas where forests are virtually absent, approximately 13 per cent of tribal household income is derived from forests (Yadama *et al.* 1997). NTFPs play a critical role in the livelihood strategies of tribal households as poverty increases (Ramamani 1988; Hegde *et al.* 1996; Godoy *et al.* 1995). In essence, the stakes are high, and there is a need to ensure the involvement of those who are primary users of forests. Any intervention involving tribal households must also recognise competing claims and the contentious history of tribal communities and the state vis-à-vis forests.

3. Conflict over forests

Conflicts over the forests in India are both numerous and historical. The origins of present-day conflicts over the forests can be traced to the Forest Act of 1878, which increased the Forest Department's power to regulate and extract resources, and sanction transgressors (Gadgil and Guha 1993). Not surprisingly, resistance to state monopoly over forests was most immediate among the tribal populations. The hill tracts of Rampa and Gudem in Andhra Pradesh, where JFM is being implemented, witnessed a series of rebellions or *fituris* against colonial authority and intrusion into forests.[2] These rebellions were, in large part, a response to restrictions on forest use, production of traditional liquor and the practice of shifting cultivation (*podu*) in the forests. Further fuelling this conflict between tribals and the colonial government was the active promotion of non-tribal traders in developing a marketable trade of forest products, including palm liquor. This remains the source of conflict between tribals and a very large and powerful group of non-tribal traders and moneylenders.

In the hill tracts of Andhra Pradesh (see Map 6.1), these conflicts have been exacerbated as a result of large development projects. With the building of hydroelectric projects, irrigation dams and mining operations, many thousands of tribals have been displaced from their lands. Consequently, these landless tribals migrated from the neighbouring state of Orissa into the tribal hill tracts of Andhra Pradesh. Many tribals were not only faced with displacement and forced migration, but also a severe shortage of revenue lands – cleared of forests – suitable for cultivation. Migrating tribals, left with very few choices, settled in reserved forests, clearing the forests for agriculture and further exacerbating the conflict with the forest department.

In the tribal tracts of Vishakapatnam, existing conflicts have escalated because of ambiguities in the demarcation of revenue and forest boundaries. Such ambiguities are a source of contention as the state refuses to grant title deeds to many tribal families. At the same time, there has been a steady transfer of title deeds from tribals to non-tribals, in contravention of the Land Transfer Regulation Act 1970. Absentee landlords abound, and eviction of tribals by middlemen and contractors holding false title deeds is not uncommon. Tribal communities in this part of Andhra Pradesh find themselves battling the Forest Department on the one hand, the Revenue authorities on another. Pressures on tribal life and resources mount with the steady influx of petty traders, contractors and middlemen with contempt for the tribal way of life.

Map 6.1   Study area: Tribal hill tracts, Vishakapatnam District, Andhra Pradesh, India

Note: The high altitude tribal area zone lies between 17.13–19°.09 N and 80.22–84°.33 E. The area spans Srikakulam, Vizianagaram, Vishakapatnam and East Godavari districts.

Added to this list of competing claims on forests and revenue lands in the tribal hill tracts are the mining interests of large industrial houses of India. In addition, the resurgence of the Naxal movement – an extreme left movement – in the tribal regions of Andhra Pradesh only increases the volatility of the political, social, ecological, and economic conditions in tribal communities.

The upshot of all these competing claims has been a steady erosion of trust in the tribal communities. It is not uncommon to see tribal villages marred by internal distrust as well as distrust of outsiders. In this cauldron of conflicts and distrust the Forest Department launched their JFM programme to forge collaborative arrangements for governing forests with tribal communities.

4. The Joint Forest Management (JFM) programme

In India, tribal perspectives on forests and government policies affecting forests have historically followed very different trajectories. State policies driven by industrial demands were in direct conflict with the customary use of forests by forest-dwelling tribal populations. In the last decade, various state governments have pursued an experiment to devolve authority for forest management to local communities through the JFM programme. This new approach to control, protection and management of forests has profound implications for forest dwelling tribals.

Historically, Indian forest policies have alienated people from the forests. This has heightened the rates of deforestation. Post-independence forest policies contributed to an expansion in agricultural production met industrial demand for raw materials, and tightened control of forest lands through restricted access to forests and forest products (Haeuber 1993). Forest protection policies increased the hardships of vulnerable social groups by denying them access to forests (Barraclough and Ghimire 1995). While the state took responsibility for managing forest resources, it did not have the commensurate resources to manage and police the forests effectively. Before state intervention, forests were managed as communal property; the crucial role of forests in the economic subsistence of individuals, families, and community was the basis for such management (Chopra *et al.* 1990). A failure to recognise community control of forests led to a collapse in institutional norms that were instrumental in protecting and managing forest resources for local use. A shift in property rights to the state steadily undermined the rights of tribals to use and extract forest resources. Displacement of property rights further impoverished the tribal household, whose survival is linked to the availability and access to forest resources.

Beginning with the National Forest Policy of 1988 to the current JFM policy, the thrust of Indian forest policy has shifted towards forging management partnerships with local communities. In rediscovering a legitimate role for local communities in self-governance of forests, the state has begun to devolve control of forests. The 1988 National Forest Policy, in large part, was the beginning of a new initiative that delegates responsibility for managing forests to local communities. National forest policy in tandem with a circular issued by the Government of India in 1990 serves to: legitimise local communities' access to forests; encourage communities to form forest management committees; and guarantee a portion of the produce from the forests (Singh and Khare 1993). This new policy proposal – Joint Forest Management – promotes a partnership between the state and local communities to manage forests for the benefit of the people and the state. Co-management links forest protection and the livelihood strategies of people dependent on forests.

According to some estimates, 15,000 forest protection committees (FPCs) are currently operating in India. The range of these FPCs extends across southern Bihar, West Bengal, Orissa and the northern tribal regions of Andhra Pradesh (Poffenberger 1995). Non-governmental organisations (NGOs) have been playing an important role in facilitating this new partnership between forest departments and local communities. NGOs have been instrumental in articulating the needs of forest-dependent communities to the state. Under JFM in Andhra Pradesh, households in a village or a cluster of villages have the right to become members of a forest protection group. A management committee representing the forest protection group is constituted to implement the JFM plan. The managing committee has a two-year term. A management committee is composed of 10–15 elected representatives from member forest protection committees. Women should make up 30 per cent of the committee. The committee also consists of a forest guard, representatives from NGOs, village officials, a representative of the tribal development authority in the case of a tribal area, and a deputy range officer.

Responsibilities of a FPC include protection against grazing, fires and thefts of forest produce, development of forests in accordance with the management plan, and provision of assistance to forest officers in the development of forests (SPWD 1993). A Vana Samrakshana Samithi (VSS), or forest protection committee, has usufruct rights to non-reserved items. Non-reserved items are leaf and grass fodder, thatch grass, broom grass, thorny fencing material from specified species, fallen lops, tops, and twigs used for fuel. A VSS does not have automatic right to products classified as reserved items such as *tendu* leaves that have been previously leased. After three years of its formation, it is entitled to 25 per cent of timber and poles harvested. The Forest Department can sell any unused portion of the 25 per cent and all of the resulting revenues are to be given to the forest protection committee. A forest protection committee is also entitled to a third of the revenue from the 75 per cent share of the forest department.

JFM represents a paradigmatic shift from previous models of development forestry. The emphasis here is on developing trust and reciprocity within forest communities and between forest department and local communities. Ideally, this is done through a set of carefully crafted institutional arrangements for the governance and sharing of forest resources. Such institutional arrangements must simultaneously hold the community and the Forest Department responsible and accountable to each other. A fundamental aim of such arrangements is 'to break the barrier of mistrust which divides state agencies and user groups' (Vira 1997). This is necessary so 'that local knowledge of the resource and existing social structures can be used to develop more effective strategies for resource use' (Vira 1997).

We should like to turn our attention to the theoretical underpinnings of resource co-management arrangements. We will argue that norms, trust and information channels supplied via civic associations, that is, social capital, are critical in explaining variations in the supply of public goods and common assets (community forests) across communities. What follows is a theoretical discussion of social capital.

5. Social capital: an operational definition

Social capital is broadly understood as the complex of obligations, expectations, information provision, associations, norms, conventions, rules and trust that is embedded in the relations between members of the community (Coleman 1990). It is a public good that benefits all the members of a community – those who contribute to it, and those who do not. The social capital perspective rejects the notion that society is merely an aggregate of individual actions (Coleman 1990). It is *social* precisely because it is produced via the patterns of interaction between the members of a community. It is a form of *capital* because it

is productive (Ostrom 1996). The social capital produced via the interaction between members of a community in turn becomes a resource
for individual members to draw from (Coleman 1990). This productive
capacity affords social capital the ability to serve as an instrument of
collective action. Indeed, the simplest understanding of the concept
social capital is 'the ability of people to work together for common purposes in groups and organizations' (Fukuyama 1995). That is, social
capital can be thought of as the ability of people to overcome dilemmas
of collective action. Thus, where collective decisions and actions are
called for, they are more likely to be successfully undertaken in groups
where social capital is high. Social capital is comprised of three general
components: norms (conventions and informal institutions); trust;
and information channels (Ostrom 1990; Knight 1992; North 1990;
Putnam 1993).

Social capital is a public good. Characterised by indivisibility and non-
excludability, public goods are often undermined by provision difficulties (Taylor 1982). Therefore, public goods often suffer from the wiles of
those who would just as soon 'ride free' (Olson 1965) as participate in
the work of provision. Accordingly, the forms of social capital – shared
norms, trust and information – are generally undervalued, and therefore
undersupplied (Putnam 1993b). Undersupply is, in large part, because
direct supply of these forms is infrequent. That is to say, individual
members of a community seldom invest consciously – directly – in the
formation of either shared norms, social trust or the dissemination of
information.[3]

Given the inherent supply problems, social capital must therefore be
supplied as 'as a by-product of other social activities' (Putnam 1993b,
170). This occurs in two distinct ways. First, social capital is created as a
by-product of appropriable social organisation. That is, through networks of civic associations. Looking at trends from the General Social
Survey data over the last several decades, Putnam (1995) found a correla-
tion between trust and association membership. That is, civic networks
are a primary vehicle for the building of social trust. These networks of
civic engagement are the array of associations – tribal youth clubs,
mahila mandals (women's groups), thrift groups, etc. – which, in many
ways, engage in convention- and institution-building activity. Such
associations, while formed to further one set of goals, can aid in the
accomplishment of others (Coleman 1988). Because social capital is
often dependent upon by-product supply, much of the literature in this
area emphasises associational membership – or civic engagement – as a
proxy for social capital (Putnam 1993b).

Second, social capital in the form of institutional norms can be said to be a by-product resulting from conflicts over the distribution of resources (Knight 1992). For example, the herdsmen in Hardin's example may negotiate with one another to devise rules for use of the grazing commons. The rules themselves represent social capital in the form of institutional norms. By constructing such rules, the herdsmen can insure that distribution of the resource is acceptable. Furthermore, by negotiating rules for the use of the grazing commons, individual herdsmen can be assured that the resource itself will not be over-utilised – as in the case of Hardin's 'tragedy of the commons'.

Ostrom (1994) offers an example of this type of by-product supply of social capital. In the article, Ostrom discusses the decision-making processes that guide the development of institutions for self-governance concerning an irrigation system (including the governance of both construction and maintenance). In the particular case Ostrom discusses, those institutions for self-governance take the form of a farmer's association. Ostrom (1994: 558–9) concludes that '[t]he investment in social capital frequently takes the forms of bargaining over which rules will be adopted to allocate benefits and costs of collective action'. In sum, conflict over the distribution of resources is an important source of social capital as a by-product.

6. The making and unmaking of social capital

In light of this discussion of social capital, we will now turn to a field perspective on forging co-management strategies between the forest department and tribal communities. Our perspective on prospects and barriers to successful co-management of forests comes from the experiences of Samata, an activist non-governmental organisation (NGO), working in the tribal villages of Andhra Pradesh for the last eight years. In establishing forest protection committees, Samata was sought as an intermediary between the forest department and tribal villages. All observations are drawn from field visits to Malevalasa, Panasavalasa, Karakavalasa, Dingriput, Vanakachinta, Nandigaruvu and Nimmalapadu villages in the tribal region of Vishakapatnam, Andhra Pradesh. Samata has been active in all of these villages. For analytical clarity, discussion is organised around three phases of Samata's work in tribal communities.

6.1 Making social capital

To understand the resurgence of trust, reciprocity and consolidation of collective action in many of the study villages, it is important to focus on how Samata tackled the core problems of tribal communities. Land alienation is a fundamental problem confronting many of the households in these villages. But, tribal households are saddled with conflicts on other fronts as well. Forests surrounding many of these same villages that Samata has been working in are under pressure from mining. These and other pressures take place against a backdrop of ever-dwindling forests.

In many of the tribal villages, Samata first tackled land conflicts. Their approach was to address land grabbing and title deed disputes one tribal household at a time. This meant filing cases on behalf of an individual tribal, getting land surveys done by the revenue department, and acting as liaison to the District Collector and the staff of Integrated Tribal Development Authority. The intent of this approach was to mobilise courts – local and state – and government machinery to work on behalf of each tribal household. Samata's work entailed continual interaction with people in the villages on the one hand, and with state functionaries on the other. These efforts in pursuit of land titles on behalf of several tribal households at a time involved very intense interaction with tribal villages, and with different branches of the government, at the local, regional and state level. Accordingly, Samata was soon organising visits from middle level bureaucrats from various state agencies to interact and discuss land issues with tribals. Such visits gave the tribals a greater sense of representation, even if they did not resolve all ambiguities in land titles and records. In many ways, these interactions also provided district and regional level bureaucrats a renewed sense of purpose and accomplishment. Samata quickly gained in credibility and trust among the tribals and the state. Repeated interactions created new channels of information: among tribals, between tribals and NGO, and the state and tribals. Suddenly there was intense interaction between tribals, the NGO and government officials around the resolution of long-pending land disputes. Tribals, through this process, were better connected with state bureaucracies, not as trespassers and transgressors, but as citizens rightfully demanding a resolution to their land disputes.

Samata's credibility and people's trust in the leadership and staff steadily increased, as they were successful in tackling the most fundamental problem – namely, access to productive assets. Samata's staff had always felt and portrayed themselves as a tribal cooperative NGO. This meant that tribals themselves felt a sense of efficacy and accomplishment. Credible commitment to collective tasks in many of these villages was no longer difficult. For instance, it was now possible for Samata to organise tribal villages to attend rallies at local police stations to protest against arrests of tribals on trumped-up charges. Samata also used this renewed trust and reciprocity – social capital – to start 120 tribal women's thrift societies helping many villages to raise small pools of financial capital. These same thrift societies also took up value-addition and processing of NTFPs. Through the power of the women's thrift societies, Samata began advocating for direct marketing of NTFPs in the open market and circumventing the Girijan Cooperative Commission (which has a monopoly on all forest products gathered by tribal households). While Samata's work on land disputes primarily involved working with tribal men, the thrift societies brought many of the tribal women into Samata's social change strategies. These efforts, overall, resulted in an increase in the supply of social capital and the propensity to engage in collective action. In sum, the most important gains took place at 'the level of social and political development' (Hadenius and Uggla 1996).

A surge of activity in tribal communities, instigated by Samata, had not gone unnoticed by the forest department and the Peoples War Group (PWG), a radical insurgent group active in the Eastern Ghats of Andhra Pradesh. The Forest Department, on the verge of implementing their JFM programme, underwritten with loans from the World Bank, sought to rely on the trust and reciprocity between Samata and tribal communities. Having earned the goodwill and trust of the people, Samata was a natural intermediary to aid the forest department in the implementation of village forest protection committees in the tribal hill tracts of Vishakapatnam district.

6.2 Consolidation of social capital

In the following two years, the tribal villages and forest department accomplished a string of forest protection committees through Samata's mediation. Samata's credibility was helpful in linking the governmental apparatus with tribal villages to resolve forest boundary disputes, and other initial hurdles to mounting a co-management strategy. A major by-product of Samata's initial work in resolving land titles was the building up of norms of reciprocity and trust within and between tribal communities. Social capital produced as a by-product of land dispute resolutions in the previous years was critical in helping forge forest co-management strategies. Samata was able to facilitate a dialogue between tribal communities and the forest department to begin a series of joint forest management arrangements.

The formation of a forest protection committee in Sovva *panchayat* offers a good illustration. In contrast to other tribal areas, the people of

Sovva engage in few forest-based activities due to an absence of forests. They face extreme shortages of fodder, fuelwood and timber. Women walk distances of 10–12 km twice a week for firewood. A common source of fuel in this region is the dung cakes prepared by women. This is unusual in the tribal regions of the Eastern Ghats. Livestock is taken to hills in the neighbouring region for fodder. To obtain construction timber, the tribals travel up to 120 km. To get grass for thatching, they travel to the neighbouring state of Orissa – a day's walk away. For medicinal plants and herbs that are widely used in the villages, medicine men travel for at least two months in the forests more than 100 km away.

Five villages in this *panchayat*, on their own initiative, came together to form a forest protection committee. They began protecting 2,000 acres of highly degraded reserve forests. Forest guards were paid in kind by the villagers to monitor these degraded lands. The Forest Department, however, decided to clear whatever remained of the forest to raise a plantation. Tribals in these five villages then approached Samata for support. Samata negotiated with the Forest Department to incorporate the protected area into the JFM programme. The five villages were then registered as a recognised forest protection committee, and the village guards were retained and continued to be paid in kind by the five villages. While such a resolution of the situation is seemingly quick and simple, it could have escalated into a major conflict if the people did not trust the NGO to mediate on their behalf. A dialogue with the Forest Department, predicated on trust in the mediating NGO was instrumental in forming a forest protection committee. Otherwise, these tribal villages were sceptical of registering their forest protection committee officially with the forest department.

Another example illustrates how Samata as an intermediary organisation enabled the joint production of a forest protection committee. One of the villages, Vanakachinta, was an illegal settlement in the reserved forests. Tribals in this village were originally from the neighbouring state of Orissa, and settled here as they were displaced by construction of dams and mining. Since they settled in the reserved forests, they were continually being accused of encroachment by the Forest Department and threatened with eviction. Again Samata and the village worked with the Forest Department. A decision to register a forest protection committee and recognise the settlement as legal was negotiated. This was a clever solution, as the Forest Department then had the villagers on its side to protect and regenerate the forests. Moreover, villagers no longer had to worry about eviction after so many years of settling. Similar arrangements were also forged in another primitive tribal village where the households were encroaching on reserved forests. Tribal households in this village got a fresh lease of life by registering as a forest protection committee, and becoming jointly responsible for that forest. Examples of such synergy with the help of Samata abound in this region.

In one case, where Samata was not involved at the outset, difficulties ensued. In Dingriput village, the Forest Department initially formed a forest protection committee. Problems soon began to arise because of an ongoing forest boundary dispute with a neighbouring village. In their fervour to start a forest protection committee in Dingriput, the Forest Department had failed to consult a neighbouring village. Samata had to mediate the boundary disputes between the two villages before the forest department could enter a joint forest management arrangement.

It should be noted that the above examples only make a case for how tribal communities and one NGO are able to make credible commitments, enabling several tribal communities to enter into JFM partnerships with the forest department. It is still very difficult to envisage, in this conflict-ridden region, the Forest Department and tribal communities making lasting credible commitments without the help of a mediating NGO. Tribal villages, more so than the Forest Department, have to rely on an NGO to enable joint partnerships with the state. These commitments arise out of trust and norms of reciprocity previously established with the NGO. State–tribal synergy in the above examples are entirely dependent on built up social capital – endowments – and do not give much hope for the constructability of social capital. There is an over-reliance on intermediary organisations, and in their absence, it is conceivable that the tribals and the forest department are unable to co-manage forests. Just such a scenario has emerged in these villages that threatens to undermine a decade of investments in trust, reciprocity and reputations that have been beneficial to tribal communities.

6.3 Unmaking social capital

A series of events have unfolded resulting in Samata leaving the tribal hill tracts, and severely undermining JFM arrangements between the Forest Department and the tribal communities. The events that precipitated this situation have to do with powerful mining interests on the one hand, and the People's War Group, an armed left movement on the other. Samata had been engaged in a public interest litigation to stop Andhra Pradesh from illegal leasing of lands in scheduled areas to non-tribals for mining. Birla Pericalase leased 120 acres of land near Nimmalapadu village to mine calcite. The state government commissioned the Border

Roads Organization to build 20 km of road from the mine site to the nearest all-weather road. Such a mining lease is in contravention to the Andhra Pradesh Scheduled Area Land Transfer Regulation Act 1959. Moreover, the access road was being built through 6 km of reserved forest which is in violation of the Forest Conservation Act 1988 and the Environment Protection Act 1986. The estimated cost of the road was US$4 million, and projected revenue to the state was US$2 million per year from this project. Birla Pericalase itself expected to earn US$57 million from the mining operation. Samata filed a public interest litigation against the State of Andhra Pradesh in the Supreme Court challenging the mining leases granted to non-tribals in a Scheduled area.

In tandem with the public interest litigation, Samata also began organising tribal villages along the planned 20 km road to the mine site in Nimmalapadu village. The strategy was to mobilise tribals to oppose any road building through their villages. Birla Pericalase was intent on completing this road, and to this end they were offering to purchase tribal lands at Indian Rs. 1,500 per acre. Samata was mobilising on the one principle that it is better to own productive assets such as land than to be displaced to work as wage labour for a mining concern. Many tribal households along the planned road held firm against selling their land to the mining company. This only prompted Birla Pericalase to increase the going rate for an acre of land from Indian Rs. 1,500–150,000. Tribal households held strong as the public interest litigation was winding through the judicial system. Samata's rising importance and central role in this struggle had an unexpected outcome – the entry of the Peoples' War Group (PWG, also referred to as Naxals) into the fray. PWG has been experiencing a decline in its power in the region, especially with the rise of NGOs providing legitimate means of recourse and social justice.

The entry of PWG is best recounted in the words of a member of Samata:[4]

> I'd only like to add that the extremists [PWG] have come, *en masse*, to Paderu region and have suddenly become very active, especially in Nimmalapadu area. They are holding village meetings and have issued paper statements in almost the same phraseology as we have been talking in the last five years. This creates serious problems/twists to Samata's work and we have to tread very cautiously. There is a strict police vigil (Naxals have taken away explosives from the BRO [Border Roads Organisation] camps and left threatening messages to BRO [Birla officials]) on Samata also. This makes it quite impossible to

move around in the area, hold meetings or organise people. We may have to restrict our work, for the time being at least, to legal and media advocacy. The people of Nimmalapadu, on the other hand, have told the Naxals that they'd rather carry on their fight with Samata than join them [PWG].

PWG failed in an attempt to refashion itself as a champion of tribals and assume a leading role in the mining conflict, signalling their marginal role in the development of tribal villages. They persisted nevertheless in calling village meetings and talking about the mining conflict.

In July 1997 the Supreme Court of India issued a judgment in favour of the tribals and Samata. The judgment was clear in stating that scheduled areas cannot be leased to non-tribals, that the government does not have the right to lease tribal lands, and that the government cannot lease lands in scheduled areas for mining as it is in violation of the Land Transfer Regulation Act 1959, the Forest Conservation Act 1988 and the Environment Protection Act 1986. The judgment was far-reaching in that it not only pertains to Andhra Pradesh, but also to states with similar laws. This victory further consolidated Samata's presence in this part of the tribal region. On the one hand, this experience with Samata increased the faith of tribals in the judicial system. On the other, it eroded the latent support for radical groups such as the Peoples' War Group. PWG has been very threatened by NGO activity in the tribal region for sometime. The Supreme Court judgment considerably weakened their support base, and pushed PWG to take desperate measures to reclaim their place as the sole voice of oppressed tribals. PWG, feeling threatened and sensing a severe loss of support among tribals, declared the entire Eastern Ghats region a 'guerrilla zone'. Claiming NGOs in the region to be puppets of multilateral and bilateral donor agencies, the PWG issued orders for Samata and other NGOs to leave the region or face execution. Samata was forced to pull out of the region, virtually halting their day-to-day work in the villages.

The government of Andhra Pradesh has been noticeably silent, and has failed to take any measures against the PWG. The government of Andhra Pradesh in its silence has severely undermined any chance of co-management of forests taking hold for the long term. The Forest Department has also failed to consider the implications of the rise of PWG on its joint forest management programme. The Forest Department views JFM as another project. There is no recognition that JFM ideally should constitute a shift in the way people and the state relate vis-à-vis forests. In this sense the state has failed to consider the social, cultural and the political dimensions of co-management. Apart from the rhetoric of participation, the Forest Department has not changed the way it perceives tribal communities. New barriers of mistrust have emerged with the departure of Samata and the re-emergence of PWG. In the current environment, trust, reciprocity and information flow between the state and tribal villages is difficult to produce and even harder to sustain. How the events described above will affect state–community partnerships in co-managing forests is uncertain.

7. Conclusion

For sustained participation of tribal communities in co-management of forests, we argue, it is necessary to focus on the dynamic between macro social, economic and political forces and micro-institutional JFM strategies. To this end, we have applied the emerging discourse on the importance of social capital for co-production of public goods to the case of co-management of forests by tribal communities and the state. While social capital is essential for co-management strategies, sustaining and consolidating social capital is complex. Social capital is the raw material of collective action that is essential for initiating, nurturing and sustaining a forest protection committee. The Samata case demonstrates the fragility of working with communities in volatile environments, and also illustrates the difficulties of building trust and norms of reciprocity in tribal communities. It only takes a few radical elements to undo a decade of deliberate and methodical development work. There remain many threats to social capital and the supply of collective action in this region, as it is plagued by non-tribal exploitation of tribals, conflicts between the Forest Department and forest dwelling tribal communities, and the ominous presence of the People's War Group.

Notes

1. These products include tamarind (*Tamarindus indica*), adda leaf (*Bauhinia vahlii*), gum karaya (*Sterculia urens*), marking nut (*Myrobalanus chebula*), mahua flowers and seeds (*Madhuca indica*), wild brooms (*Thysanoloena maxima*) and soap nuts (*Sapindus emarginatus*).
2. The first rebellion known as Rampa Fituri was in 1803. The second revolt was in 1879 against the Muttadars and the British, and this rebellion was wide spread. Then from 1922 to 1924 was a third rebellion of Koya tribals instigated by Alluri Seetharama Raju. The fourth rebellion by the Naxalites was between 1968 and 1972.

3. Clearly individuals maintain norms of behaviour, personal trust and participate in information dissemination. The distinction is here made that individuals do not directly participate in the establishment of shared norms,
social trust or the informing of the community. The latter three cannot be
undertaken by the individual alone; they are products of the community as a
whole.

4. Personal communication with a member of Samata, 1997.

Part III

Attitudes and Responses of Intervention Agents

7

Learning from Participatory Environmental Impact Assessment of Community-centred Development: The Oxfam Experience

Koos Neefjes

1. Introduction[1]

This chapter looks at lessons learned in the field of participatory natural resources management by a relatively large non-governmental development organisation, Oxfam GB[2] and some of its national counterparts. It describes a methodological approach to assessment of environmental impacts of community-centred development and elaborates on the impact of adopting this approach on local people's livelihoods and environments and on organisational capacities and skills of staff and counterpart staff. It concludes with some assertions about the implications of both 'participation' and a poverty focus, which constitutes Oxfam's core mandate. The chapter presents a picture of 'learning-in-progress' and draws on internal data from sometimes sketchy reviews and documentation of projects, wider programmes and of training efforts.

Oxfam promotes a kind of development that is strongly 'rights-based' and can be characterised by current keywords such as holistic, empowerment, participation, structural, socially just, gender aware, sustainable, and so on. It does not just aim to effect changes in a number of communities across the globe; its development approach is people- and community-focused, and it also wants to ensure that (macro) policies change in order to increase poor people's opportunities to help themselves. Methodologies that are employed in 'strategic' (long-term) planning of wider development programmes and project planning, mon-itoring and evaluation must be consistent with goals of empowerment and sustainability and with the principles of social justice and involvement of beneficiaries. 'Participation' is the buzzword here, and has been shown to be a challenging notion when taken to mean involvement of stakeholders with conflicting interests. Oxfam and its local counterparts wish to be learning organisations, dynamic organisations that respond to changing realities and contexts, and they do not want to adopt dogmas or simple truisms about, for example, what is 'green' or environmentally correct. 'Green' development and natural resource conservation cannot be pursued for the environment's sake alone, and it cannot be taken for granted that environmental regeneration and protection are synergistic with poverty alleviation.[3]

In the next section a brief explanation is given of the concept of 'sustainable livelihoods', an instrument for analysing changes in people's lives and environments. Building on this, section 3 presents a methodological approach to ensure environmentally sustainable development, which Oxfam has adopted in much of its international development programme. Section 4 introduces achievements and impacts on staff and organisational capacities and section 5 discusses assessment of the impact of projects and programmes on natural resources. Section 6 takes up the debate about whether participation and sustainable natural resource management (SNRM) are compatible, while section 7 concludes with some thoughts about future prospects for research and learning.

2. Conceptualising environmentally sustainable development

According to many Oxfam partners and staff the meaning of *environment* (and the meaning of *sustainable development*) must include social relationships and infrastructure and should not be equated with 'natural resources' (Stubbs and MacDonald 1993). Indeed, from 1992 Oxfam has formulated 'achieving sustainable livelihoods' as an organisational objective which has helped to view the quality of and access to environmental resources in relation to poverty (Oxfam 1992). This phrase was used loosely in setting general aims, covering the breadth of activities of the organisation including 'fair trade', 'income generation' and primary production for subsistence. Nevertheless, the concept was presented in literature as an aid for establishing conceptual links between environmental quality, people's rights to resources and capabilities to manage resources. A need for an analytical framework was strongly felt amongst development professionals, in particular in NGOs where assumptions prevailed that poverty alleviation, empowerment and participation of local and 'environmental care' would usually go hand in hand. The sustainable livelihoods framework builds strongly on academic work and preparations for UNCED from the 1980s and early 1990s.[4]

Essentially, a livelihood is defined by Oxfam as *a means of living, not just of production*, and sustainability is defined as *the ability to maintain and improve livelihoods while maintaining or enhancing the global assets and capabilities on which livelihoods depend*. Oxfam built on the early literature on sustainable livelihoods and adapted it for practical purposes in 'strategic' planning and project management. It asserted that livelihoods can become more sustainable and vulnerability of poor people can be reduced if the following perspectives are incorporated in, and shown by the analysis:

1. *assets* in the local environment (for example, land, water, services and infrastructure, money, forests) are of good and improving *quality*;
2. poor people (women, men, certain castes or ethnic groups) have access to these assets, i.e. have *claims and relative control* over them;
3. people have sufficient *capabilities* to use the assets and create livelihood opportunities;
4. people have the *ability to offset risks and cope with shocks*, such as natural disasters, epidemics, political violence and sudden market changes;
5. changes imply *positive contributions to the natural resource base* (e.g. land, water and biological resources), the *human built environment* and *social resources* such as health services, and they are neutral or *contribute positively to other livelihoods*;
6. *sustainability* is understood in an economic, environmental *and* social/institutional sense: the economic or financial returns, the changes in environmental quality, or the social benefits in terms of health taken in isolation cannot show the success of programmes.

This framework has several limitations, for example, it does not provide criteria for what is environmentally or socially good or bad; it merely helps to ask some important questions. The framework is not a theory and thus does not explain failure or success. It is not clear which social unit or groups need to be central to the analysis, given that impacts on (natural) resources are usually different for different social 'actors' and that social relations are affected by environmental change. The analytical framework of itself does not 'demand' participation of stakeholders in the analysis. It also does not offer operational support at project level or in analysing and influencing institutions and policies. Nevertheless, it can be seen as an instrument for developing 'strategic plans', setting objectives and improving the understanding of linkages between environmental change and poverty (alleviation) at the micro-level, as the next section argues.

3. Methodologies for environmentally sustainable development

3.1 Strategic planning

At programme level Oxfam routinely develops 'strategic plans' which set out a contextual analysis and formulate medium to long-term aims and activities. These plans exist for country programmes regional programmes and for Oxfam's entire international programme,[5] and some counterpart organisations have adopted similar procedures. Logical frameworks are a central feature of strategic plans, in which aims, objectives and indicators are made explicit. They are written on the basis of a contextual analysis, programme strengths and weaknesses, and so on. Strategic planning is essentially an internal process to Oxfam but various 'stakeholders' are consulted, especially Oxfam's counterpart organisations. The strategic planning process has helped Oxfam's programmes to formulate the changes in poverty and livelihoods that it pursues in specific contexts, including changes in environmental quality and people's access to natural resources.

A review of the first strategic plans of Oxfam's country programmes, brief desk studies and anecdotal field observations in 1992 demonstrated that a number of key issues needed to be addressed. Although programmes often impacted positively on natural resources, for example through promoting sustainable agriculture or soil and water conservation measures, environmental change and impact had not always been made explicit. Moreover, in the majority of Oxfam programmes environmental degradation and poverty (Oxfam's core aim) were not linked conceptually. Notable exceptions to the latter problem included the Philippines programme where practical regeneration of fishing grounds by grassroots groups, support to intermediary non-governmental organisations (NGOs) and support for national networking and campaigns on fishing rights were all undertaken with the shared goal of supporting poor fisherfolks' livelihoods and the natural resources on which they depend – synergy between these two goals had been found. Impact indicators in the country-strategic plans, that is, indicators that showed some general and lasting change and that were related to environmental resources included availability of safe drinking water; reduced malnutrition; increased availability of environmental resources; reduced resource degradation (Neefjes 1992).

These indicators illustrate the importance of natural resources, but none of them expresses the notion of access to and control over resources. Strategic planning guidelines that were developed later contain guidance on what sustainable livelihoods mean and what the relevance is in several steps of the planning process (Oxfam 1995). Strategic plans are regularly updated, and the ones that are currently used show some evidence of the use of this conceptual framework and a marked shift in some particular areas. There is now a clearer articulation of rights to environmental resources, in particular land,[6] which are a key necessity for livelihood security, especially in rural areas, whilst these rights are also key in achieving better environmental management. In the Vietnam programme, for example, support in the northern province of Lao Cai for implementation of land reform legislation has stimulated farmers to plant trees and helped reduce a local form of 'slash and burn cultivation'. Strategic plans now also conceptualise the links between national and international policy and local environmental management. In Vietnam the learning regarding land tenure at the micro level led to engagement with provincial and national authorities and contributed to a country wide shift in implementation policy (Neefjes 1998a). Following a review in 1997 of support for nine grassroots NGOs working in conservation farming in Kenya, Oxfam and partners planned to strengthen their policy analysis and national influencing capacity, which appears to have started happening, according to an update of that review (Oxfam GB 1999). However, country and regional programmes still make few links between global (international) environmental change and local poverty or local environmental change, although Oxfam does recognise the importance of these links.

3.2 Environmental impact assessment: the project level

A methodology for assessing actual environmental change at project level is known as Environmental Impact Assessment (EIA). EIA was developed to predict possible environmental damage of large-scale projects such as industrial and infrastructure development schemes. It has been incorporated in legislation and regulation in industrialised countries, a large number of developing countries and many international development agencies have adopted environmental assessment guidelines. Through the process of EIA normally a number of scenarios with different environmental impacts are assessed. The (negative) environmental impact is predicted and certain mitigation measures may be proposed. The impact and the mitigation measures are, as much as possible, expressed in monetary terms, or they are quantified in other ways in order to make comparison possible.

Conventional methods for EIA are not well suited to small-scale, community-based and participatory projects for a number of reasons. EIA is usually not participatory (being consultative at best) and therefore incompatible with community-based planning and evaluation processes; it may involve high costs (in the context of low-budget projects); it usually fails to draw out socially differentiated impacts; and moreover, it is usually meant to minimise negative environmental impact instead of enhance positive impacts on both people and their environments. Furthermore, EIA guidelines normally do not seek to support both planning and post-project impact assessment, but concentrate on the former, whilst it is the latter that can be expected to contribute most to institutional learning and improving practice. The effectiveness of conventional EIA has also been questioned, as well as the costs in the context of cheap and small-scale projects.

Some bilateral agencies have produced or are preparing EIA guidance for NGOs, either to enable them to comply with national statutory requirements or their own planning procedures, but this remains geared to large projects. Existing guides or manuals on EIA (or better: 'environmental screening') for low-budget projects are not very participatory in their approach and cover only a few of the great diversity of sectors addressed by NGOs (Maesson 1992; Hughes and Dalal-Clayton 1996). Despite all these limitations EIA and in particular environmental screening do offer checklists that may prove useful in situations where experts are not available and development workers need to be confident that a full breadth of questions about environmental resources is being asked (Eade and Williams 1995).

NGOs like Oxfam and its partners have long assumed that the negative environmental impact of their projects was negligible and indeed it was not until 1991 that Oxfam adopted a question on environmental impact in its standardised grant appraisal procedure. Policy aims on sustainable livelihoods were reflected in a one-page guideline for Oxfam staff who have to answer this question.[7] This guideline is not a checklist, but proposes six fundamental questions which reflect the sustainable livelihoods framework (see Figure 7.1).

These suggested questions were distributed widely amongst hundreds of Oxfam staff in 'field offices', in three languages, but they are not

---

When thinking about:

(A) NATURAL RESOURCES (*such as water, agricultural land, rangeland, urban land, air, forest, flora and fauna*), and

(B) HUMAN CREATED RESOURCES (*such as shelter, water supply, sanitation, health care systems, schools*)

You could ask yourself:

- **What resources are important in local livelihoods?**
- **What is the quality of those resources?**(*e.g. soil fertility, drainage potential, pasture quality, accessibility and quality of hospitals/schools, congestion, sanitary conditions etc.*)
- **Who uses which resources?**(*women, men, children, disabled, ethnic groups, social classes, etc.*)
- **Who controls decisions about how these resources are used?**
- **Who is helping to sustain local resources and who benefits from this?**
- **How will the situation be affected by the project?**

Figure 7.1 Questions reflecting the sustainable livelihoods framework

used widely. Screening of samples of hundreds of grant applications before its distribution in 1994 and also in 1999 indicate a very limited uptake where the question of 'environmental impact' is answered, even though a very large part of the grants address such questions under other headings. Indeed, Oxfam has, over the past five years, spent about 45 per cent or £140 million of its grants money on work that relates to natural resources, in particular on water supply and agriculture.[8]

3.3 Participatory appraisal, monitoring and evaluation

Participatory Rural Appraisal (PRA) (also known as Rapid Rural Appraisal or Participatory Learning & Action) is by now widely known as a methodology or approach used at project level. It is described in vast amounts of literature and strengths and weaknesses are often hotly debated.[9] With PRA tools and learning processes genuine participation of (local) people is pursued through the facilitation of dialogues ('semi-structured interviews') in which much attention is given to outsiders' behaviour, good questions and to simple analytical tools (diagrams) that enable the dialogue to focus on a particular subject. The tools, behaviour and processes of PRA can be usefully adapted to a very wide range of situations, from organisational analysis to project analysis, in early stages of projects and in later stages of projects. PRA can also produce large amounts of relevant and valuable information and insights on environmental and socio-political aspects, regarding present and past.

However, PRA can be criticised in several ways. It is not always well integrated with existing project cycles for example, although recently applications of the approach in 'participatory monitoring and evaluation' (PM&E) is being explored with renewed enthusiasm (Harkes, this volume, chapter 8). It does not offer a clear analytical process and guidance for high levels of quality and trustworthiness of data and conclusions (what happens depends very much on the lead facilitators) (Pretty 1994). Only recently has PRA become more explicit about the analysis of social differences (Mosse 1995; Guijt 1996). Furthermore, it is usually not supported by any kind of explicit analytical framework and, for example, does not address questions about the meaning of (environmental) sustainability, or the relevance of environmental degradation to poverty alleviation.

Concerns that conventional monitoring and evaluation do not always provide good insights into the actual difference that development activities make have prompted development agencies to focus more on project *impact*. Impacts can be changes of any kind, but in practice the changes that are looked at include changes in livelihoods, in people's institutions and in policies. In much current practice of impact assessment participatory approaches are stressed, but not simply the participation of a homogeneous group of 'beneficiaries'. The development (and negotiations) of impact indicators and the measurement or assessment of those indicators are also important. Impact assessment is concerned with impacts that were planned as well as changes that were not, and it has a strong analytical focus: a critical aspect of impact assessment is finding ways of actually attributing changes to development interventions. Some of this work is being given a clear process focus, conceiving longer-term processes in which multiple stakeholders operate in some kind of facilitated negotiation as regards changes, attribution of those changes to certain activities and modifications of development practice (Roche 1999; Harkes, this volume, chapter 8).

3.4   Hybridisation: participatory environmental assessment

So where, one might ask, is natural resources quality in the very broad concept of sustainable livelihoods, in PRA practice, in PM&E or in participatory impact assessment? Likewise, how can citizens and their organisations participate in formal EIA and their wishes and insights be made compatible with the setting of environmental standards and criteria? It is obvious that a hybrid form is required if all the multiple objectives are to be served; one that might be labelled participatory environmental assessment.

Oxfam has promoted such a hybrid form for project management since late 1992 in which the framework of sustainable livelihoods was put forward to look at the general context, and participatory tools are combined with environmental checklists (Eade and Williams 1995). Oxfam and also Novib have undertaken a number of workshops with staff and staff of counterpart organisations in which all these tools were applied in learning processes with citizens groups and also officials. Typically, these workshops followed the systematic process depicted in Figure 7.2.

In most cases a group of workshop 'participants' consisted of Oxfam and partner staff, and sometimes also some representatives of beneficiary communities. These people were generally of mixed background and skills and with similar numbers of women and men. The first step was to acquire a basic understanding of sustainable livelihoods and EIA-based checklists, which enabled reading of secondary documentation and formulation of research questions. Some 'dry run' practice of PRA tools and behaviour was also generally necessary. This group of

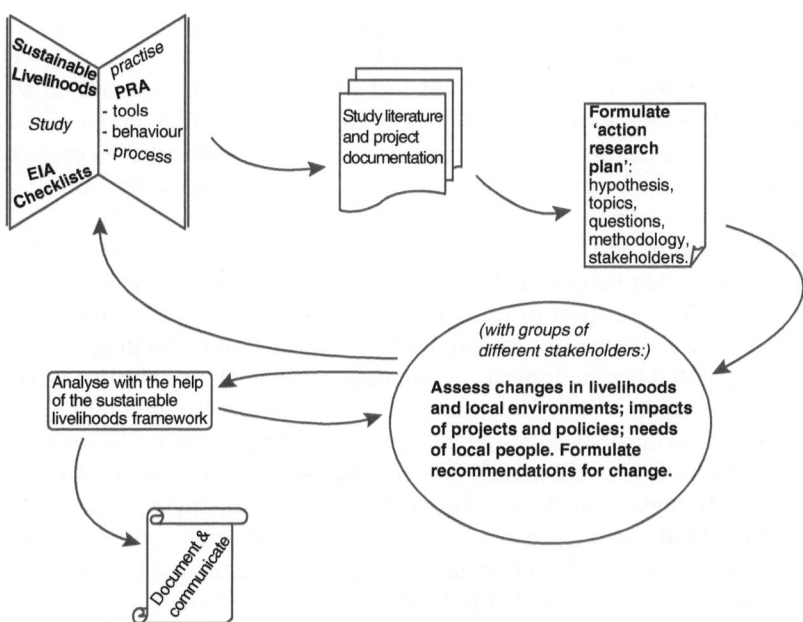

Figure 7.2   Participatory environmental assessment

people then turned into facilitators of meetings, dialogues and interviews with various stakeholder groups, for example women and men
from neighbourhoods or social classes or officials. Sometimes such
groups were brought together at a later stage to address issues of difference and consensus, all of which aimed at arriving at a good level of
shared analysis of impacts, needs and/or recommendations for future
action. The workshop participants used their PRA skills, knowledge of
secondary information and also analysis with the aid of, for example,
the sustainable livelihoods framework in order to enrich the analysis
and negotiating with the stakeholders. Some of them documented the
'findings'.[10] But has this hybrid led to results, and was training and
learning in Oxfam and other NGOs effective in improving both natural
resource quality and alleviating poverty? The next two sections
attempt to answer this question.

4. Impacts on staff and organisational capacities

Sessions on the 'hybrid' of participatory tools, conceptual frameworks
and EIA-based systematisation were organised with staff and partner staff
in/from more than a dozen countries and reports were widely distributed across the organisation. Other learning efforts have also included
aspects of sustainable livelihoods, PRA and, in a few cases, EIA. They
include staff exchanges, production and distribution of formal and informal publications (both external to Oxfam and internal[11]), and various
thematic workshops (for example, on sustainable agriculture, project
management, strategic planning and gender relationships) and on-the-
job learning through visits of national and international consultants. All
this is in addition to inputs of the 'guideline' type mentioned in section 3
under strategic planning EIA. However, Oxfam works in over 60 countries so that the impact of these efforts on the capacity of the whole of
the international programme can only be expected to be limited.

As suggested before, indications are that strategic plans have improved
over the years through, for example, focusing on environmental rights,
access and control over resources instead of simply the availability of
resources. Analysis of Oxfam's grants databases also suggests that there
are trends towards increased support for environmentally sustainable
livelihoods at project level. Such trends suggest that some capacity
improvement of the organisation and of staff of Oxfam and partners has
occurred with regard to natural resource management. However, the
array of learning-events with some bearing on natural resource management and poverty alleviation makes it virtually impossible to attribute

change to specific efforts. Nevertheless, anecdotal feedback and observation, a small number of in-depth interviews with former workshop participants and a survey of opinions about the longer term effects of the 'hybrid' workshops does suggest certain impacts.

The sustainable livelihoods framework seems most useful in the formulation of aims and general policies of programmes, enabling them to include environmentally sustainable goals within a poverty alleviation framework (as happens in 'strategic plans', in which higher-level managers have a strong input). 'Front-line staff' very rarely use the framework in project appraisal or assessment. In most cases planning and analysis happens in 'sectors' – agriculture, forestry, income generation, and so on. The sustainable livelihoods framework is too abstract for these members of staff. Nevertheless, many appreciate the idea of 'sustainable' as something they learned about in the workshops, although not necessarily in an environmental sense. EIA-based screening checklists are hardly used at all at project level.

Uptake of PRA was more successful. Several participants of the workshops on the 'hybrid' approach feel that the PRA approach has not yet been applied sufficiently in routine monitoring, and in some cases participants felt that PLA/PRA tools are used mechanically. There is also limited confidence to conduct the analytical process associated with PRA and indeed PEA. Despite these shortcomings, the survey indicates that behavioural changes are an important gain for the participants and several PRA tools are used in different situations of 'development practitioners'. This is seen as a contributing factor to improved levels of trust and better communication between project workers and local people. Individual participants also shared their new skills and ideas with colleagues, in particular those related to PRA, and claim a fair amount of learning by others.

The different assessments in Oxfam also suggest that follow-up to workshops is essential for achieving real uptake of the messages and shared learning and lasting change in behaviour. Adoption of new technologies and practices is in any case a very slow process. For example, local counterparts of Novib in Uganda were trained and guided in a range of sustainable agriculture practices and principles of participatory technology development, and they had received PRA and other training too. However, after several follow-up workshops the actual uptake by staff and by farmers was limited and a large number of training needs remained unfulfilled (Neefjes and Nakacwe 1996).

Some staff in Oxfam's country programmes indicate that they 'always' use the one-page guideline regarding environmental impact when developing grant proposals for approval by Oxfam, whilst others have used it occasionally or for a short while. Yet, observation suggests that this happens only in a minority of Oxfam offices and a random selection of approved grants has shown that the suggestions from the guideline are rarely followed strictly. The distribution of the guideline (in several languages) has had limited impact at best.

In conclusion it can be said that there is some evidence for staff learning and impact of the specific workshops, but it is limited. A key factor with all capacity-building efforts is a very high staff turnover, both in Oxfam and in local counterpart organisations. In some cases where the larger training efforts in the 'hybrid' were organised more than half of the original trainees are no longer with Oxfam or counterparts. This fact calls for structured (and elaborate) staff induction programmes and clear, written documentation on the methodologies outlined in this chapter, as well as elaborate and continuous staff training and development – a widely debated issue in Oxfam.

5. Impacts on livelihoods and natural resources

So what is the impact on natural resources (and socio-economic aspects) of the projects where these methodologies were used, and how does that compare with other situations? Again, hard evidence for such analyses does not exist, and would require very elaborate, costly and unjustifiable research programmes. The evidence must be found amongst anecdotal feedback and documentation from project reviews (that is, several years after the 'hybrid' was introduced), and the survey of opinions referred to above (Neefjes and Woldegiorgis 1999).

Impacts on natural resources in projects where the 'hybrid' approach was introduced are limited and always positive – in fact, in all cases the workshops led to new, more holistic and strategic initiatives, many of which related to natural resource management. These new initiatives are seen to be supported by more grounded data, especially as a result of the PRA elements of the 'hybrid'. In section 3.1 a reference has already been made to work in Vietnam that supports this. In North Tokar in Sudan broader awareness of the importance and local knowledge of bio-diversity was established in participatory work, which is extremely important as it is a situation where invasive mesquite (*Prosopsis juliflora*) is devastating both livelihood and survival opportunities (Neefjes 1998b). Good analysis of local agriculture and markets with local farmers has led to various attempts at crop diversification and intensification in Niassa, Mozambique, where negative environmental impacts from commercial cotton farming are observed (by the same peasant farmers under contracts with a large Portuguese company) (Neefjes 1998c).

Changes in natural resource quality and access to them by poorer local people, as well as broader aspects of livelihoods, have also been analysed by using the framework of sustainable livelihoods for analysis, in projects where the 'hybrid' as such was not introduced at an earlier date. Analysis of some prospective work in central Vietnam helped formulate doubts about new initiatives regarding pasture management and the introduction of goat rearing. In Omaheke in Namibia a rural water supply project was evaluated, revealing a negative impact on local vegetation as a result of water provision to cattle in already intensely grazed areas and a lack of support to the most marginalised ethnic group. These evaluations showed that the framework can indeed be practical and that it assists in systematically assessing a broad range of issues that are of immediate importance to local people and their environments. However, it is no more than a general framework, is not explanatory in the sense that it points at causes of failure and wider causal relationships. It also omits some important aspects, most notably institutional aspects and relations between project activities, markets, national agricultural policies and, for example, land tenure policies.

As with many other programmes with small and dispersed national NGOs Oxfam devotes attention to developing local contacts and networks, but the potential for influencing policy is not always realised. The Vietnam example mentioned in section 3.1, where this did happen, and the Kenya case where it has now started to happen are exceptions rather than rules. Policy changes are however central to livelihoods, natural resource management and to environmental sustainability, and they may be necessary in defining land tenure, arrangements for food marketing, or priorities for national agricultural research.

6. Participation and sustainable NRM: a contradiction or synergy?

Participation can take many meanings and forms (Adnan *et al.* 1992; Nelson and Wright 1995; Chambers 1997; Abbot and Guijt 1998). It can imply, for example, selective consultation with certain people, or complete self-mobilisation by local groups and activists. If it is assumed that power over decision-making lies with a small and narrowly defined group of people, it might be labelled consultative as far as most stakeholders are concerned. A process in which decisions are made more collectively, and are expected to be a compromise between many different stakeholders, must find ways of being more inclusive.

In the latter type of 'participatory' process problems are gradually analysed, in separate groups and together, and solutions grow from research, discussion and various interactions. The focus of research, analysis and indeed action is likely to be on certain key individuals and groups of social 'actors' (Long and Long 1992). Such processes are no longer strictly defined and structured, and decisions emerge from them, instead of being taken at fixed and predetermined moments by particular people. It is no longer obvious that technocrats prepare a set of scenarios for development, with different environmental and other expected impacts. Indicators (and, for example, environmental standards) for good development practice are no longer objectively verifiable, but they are mostly qualitative and often subjective. Researchers lose their independent observer status and become part of the process.

Oxfam and other development organisations are increasingly using such approaches in reviews and evaluations as part of community development processes, and the implications can be significant. Not being able to sample randomly and analyse statistically is likely to be unacceptable to the conventional scientist, whether social or environmental. Participatory methodologies need their own ways of increasing trustworthiness of data and conclusions, especially through cross checking information from multiple sources ('triangulation') (Pretty 1994; Thomas *et al.* 1998). When many stakeholders are involved, the aims of just one of them (such as Oxfam) about a focus on poverty alleviation and environmental care may well be diluted. This means that some natural resources may be used unsustainably and the very poorest in a community may not always benefit from the project, in particular at the earlier stages.

A participatory process in which local people are central, and in which local NGOs and authorities and possibly national and international organisations have a say, will concentrate on the use and management of *local* natural resources and possibly national environmental health and well-being. The methodological approach makes it unlikely that international environmental concerns would be considered. Poorer countries, and their local people, cannot be expected to be concerned about their contribution to global environmental problems because the dominant cause of these usually lies in high consumption in industrialised countries. Thus using a participatory methodology has implications for definitions of sustainability, especially what can be achieved in terms of environmental management at a non-local level.

Participatory processes of this sort must be a dialogue or negotiation process in which various stakeholders contribute both their interests and their knowledge and experience. Examples of short-term review processes that are nevertheless *not* like conventional external evaluations from Oxfam's programme are the assessment of impact of conservation farming in Kenya (Neefjes *et al*. 1997; Oxfam 1999) and also the review process in a large refugee settlement project in Uganda (Neefjes 1999a). Two related issues in these participatory processes are important and sensitive: first, there is a limit to the level of stakeholder involvement, which is partly dependent on the purpose of a review or 'action research' programme, and also on their sense of 'ownership' of the learning process; and second, the initiator and/or funder of the assessment (who will have a important voice at an early stage) must be recognised as being in a privileged and powerful position relative to other stakeholders.

7. Future research and learning

This chapter has attempted to summarise some of Oxfam's experience, and it shows a number of complexities in involving people in development projects as well as achieving environmental sustainability. There is no simple answer and no single methodology can solve the problem. We have to look and operate at different levels (programmes and projects, strategic and practical, policy and practice), and accept compromises if we are to achieve empowerment, poverty alleviation *and* environmental care. In addition, continued learning is needed, on the meaning of environmental sustainability and on the methodologies that can be used to achieve that.

Recent academic research and development of analytical frameworks on 'environmental entitlements' has also adopted the idea that environmental management is the outcome of negotiation processes at several levels and is mediated through different institutions, such as markets and land tenure regulations (Leach *et al*. 1997a, b; Carney 1998). Weaknesses of earlier versions of the 'sustainable livelihoods' framework include the fact that there is little institutional focus, whilst it does not 'model' the need for socially differentiated analysis (of environmental change, or of socio-economic gains). In this sense it does not aid Oxfam and counterparts in efforts to influence national and international policies. Any future developments in Oxfam and other development agencies should aim to ensure that they are informed by this recent academic work.

The outcome of internal survey and assessment that the uptake and impacts of some capacity building efforts regarding the framework and its practical application in project management is limited, which calls for renewed action and possibly different instruments for learning. This is because the need for better analysis of environmental sustainability issues remains, despite earlier gains. There also remain learning needs amongst higher level management on the rights of future generations and the interaction between environmental resources and the livelihoods of people from different socio-economic groups.

At the programme level, development agencies and national or regional authorities may want to engage with sector or strategic environmental assessment, covering sectors or geographical areas and a broad range of issues rather than the impact of a single project. Although this would inform higher level planning it will require further strengthening of partnerships and networks, in particular between researchers and authorities.

Notes

1. This chapter has benefited from comments on an earlier draft by Philip Woodhouse and Bhaskar Vira, for which many thanks.
2. In the rest of this chapter Oxfam GB will be called Oxfam, without wanting to suggest that the experience shared in this chapter is that of other members of the Oxfam International family. The one exception to this is the Novib from the Netherlands, also member of Oxfam International, whose partners in Uganda are referred to as well (Neefjes and Nakacwe 1996). Oxfam GB has had an annual income of £80–100 million in the past 5 years or so, which makes it one of the largest international development charities in the world.
3. Many largely successful cases of supporting poverty and environmental improvement or protection have been documented in Conroy and Litvinoff (1988) and under the label 'primary environmental care' Oxfam and others have also suggested synergy between people's empowerment and environmental care (Davidson and Myers, with Chakraborty 1992). However, there are various degrees of incorporating rights of future generations to environmental or natural endowments, as explained by, for example, Dobson (1999). Local people's interests do not *always* coincide with preservation of natural resources or nature: see for example Guha and Martinez-Alier (1997). Peet and Watts (1996) and other who write on 'political ecology' also assert the centrality of poverty to environmental degradation, qualifying it with explanation of the complexity of interactions between people and environments and also the fact that markets are the means through which faraway (rich) consumers affect methods of production and environmental exploitation.
4. Oxfam used in particular Chambers and Conway (1992), Chambers (1995); see also Leach *et al.* (1997a,b), Scoones (1998) and Carney (1998) for later developments. Related but separate theoretical work that can be seen as underpinning of the framework of sustainable livelihoods can be found in literature on gender, environment and development (for example, Agarwal, 1997) and political ecology (for example, Peet and Watts 1996).

05. A programme is conceived of as a cluster of local counterparts and projects that Oxfam supports, in for example a country or in a region consisting of several countries, and Oxfam also talks of its 'international programme'.
06. As witnessed by Oxfam's Strategic Plan of 1997: 'Oxfam will work to support pastoralists and peasants protect their land rights in Africa'. Palmer (1997) was published by Oxfam in an attempt to support learning on land rights of development practitioners.
07. A Note on the Environment Question in the PASF, internal mimeo. The question is 'what will the environmental impact of this project be (where relevant)?'.
08. From: Oxfam (1999) *Exchanging Livelihoods – natural resources edition*, unpublished collection of case histories.
09. Chambers (1997) summarises some of this debate and gives a large number of references.
10. Such processes have been conducted in project appraisals and reviews in for example Cambodia, Sudan, Vietnam, Mozambique, Uganda, Angola and Rwanda and are reasonable well documented. This was done mostly for Oxfam, whilst some of the work in Uganda was with counterparts of Novib: Neefjes and Nakacwa (1996).
11. Informal publications in this respect include *Exchanging Livelihoods*, a short series of now 4 volumes of write ups of case histories and experiences for the benefit of learning on sustainable livelihoods in practice; the case histories are generally written by staff and distributed widely amongst staff and also some partners, in three languages.

8
Project Success: Different Perspectives, Different Measurements

Ingvild Harkes

1. Introduction

Within Southeast Asia, the government of the Philippines has been a leader in decentralising management of natural resources to the local level. Between 1984 and 1994 more than 100 community-based resource management (CBRM) projects were undertaken. According to Pomeroy *et al.* (1996), the time, funds, and collective effort put into these projects have allowed implementers and participants to accumulate valuable knowledge in the area of CBRM. An overall evaluation by Pomeroy and Carlos (1997) revealed, however, that fewer than 20 per cent of these projects were identified as being successful. The criteria used to evaluate success were that the community organisation still existed and that at least a single project intervention was maintained after the project terminated. It may have been that the project components (alternative livelihood programmes, the installation of a protected area, or technology for increased fish production) were implemented at a time when the people were not ready for them, or that for the people the project components were not relevant, but *why* the majority of projects failed, is not clear. It is, however, not the scope of this chapter to discuss project failure, but to discuss *how* to measure failure, or, in that respect, project success. Indeed, a more in-depth study by Pomeroy *et al.* (1996) showed that while projects could be unsuccessful in the eyes of the implementers, the project participants did not necessarily perceive the projects as failures at all.

This chapter is in the form of a review and attempts to shed light on why the perceptions of project staff and beneficiaries are contradictory, and how we can revise the methodology of evaluation in order to capture the impacts of a project from both perspectives. The chapter begins to

128

explain the discrepancy in project evaluation with an illustration from the above mentioned study by Pomeroy *et al.* (1996). Then it takes a step back in time and explains why conventional development projects shifted to people-centred resource management. It explains the difference between the perspectives of the project participants and those of outsiders, and the consequences of this for perceptions of project success. How the 'inside perspective' can be measured objectively, and what should be measured and when, is discussed as well as the methodology. Finally, the chapter describes the obstacles to include the extra measurement in CBRM and co-management project evaluations.

2. Success or failure?

The in-depth analysis by Pomeroy *et al.* (1996) covered nine project sites that were part of the Central Visayas Regional Project-1 (CVRP-1) and the Coastal Environmental Programme (CEP) in the Philippines which started in 1984. The study showed that, while projects could be unsuccessful in the implementers' eyes, the project participants perceived them as largely successful. To be more precise, the analysis suggested that the community-based coastal resource management projects evaluated were successful despite partial or complete failure (or destruction by natural events) of some project objectives such as mangrove planting, artificial reefs, fish aggregating devices, and shellfish culture.

Illustrative are two cases: Calagcalag and Tiguib, in the municipality of Ayungon region, Negros Oriental. Impacts were measured on two levels: (1) the independent variables (project variables and context variables); and (2) the dependent variables (achievement of intermediate objectives and impacts on 'well-being' of the coastal ecosystem). The first set of independent variables are less relevant because they are comparable for both cases. The latter most clearly underline the argument in this chapter. Calagcalag is presented as a successful case, while Tiguib is classified as being unsuccessful. The first subset of dependent variables included both material objectives (see above) and non-material objectives such as training and institution building. The second subset (well-being of the ecosystem) included both human and 'natural' components. For a more elaborate description see Pomeroy *et al.* (1996).

2.1 Calagcalag

The artificial reefs (ARs), installed as part of the material intermediate project objectives, were ineffective and partly destroyed by typhoons, as happened with the Fish Aggregating Devices (FADs) that were installed.

However, both are being reinstalled. Mangrove reforestation was a failure because apparently an improper species was selected, as was the case with an oyster farming project. The goats introduced as alternative livelihood died or were not very useful – in fact, the very idea of goat's milk amused many respondents. From the intermediate objectives, only the seaweed culture as part of the sea-farming and the installation of a marine sanctuary were successful.

From the organisational and institutional perspective, the Fishermen's Organisation was active, but not widely supported – membership being beneficial to those who wanted access to the mangrove plantings, ARs and FADs, or to the alternative livelihood programmes. Activities concerning use rights and management efforts (community-based monitoring and enforcement), were successful in the sense that most respondents from the fisher community knew *why* the sanctuary and other regulations were installed, namely, to protect the fish habitat and breeding area. Generally, the perceived impacts (ten in total) on the ecosystem were positive, except for access, which was related to removal of fishing area by the installation of a sanctuary.

Pomeroy *et al.* (1996) suggest that it is important to note that this overall perception of positive change exists despite partial or complete failure (or destruction by natural events) of some project objectives. They write further that the early involvement of the fishing groups in the project gave them a better understanding of the difficulties that came with the introduction of new technologies. Additionally, the adaptive nature of the project, the willingness of the project staff to listen to beneficiary complaints and institute changes in implementation procedures, resulted in the participants' feeling that they, in part, influenced project success (ibid.). This was also the case in less successful projects, where an overall perception of positive change existed despite partial or complete failure of some project interventions. This is clear from the next case.

2.2 Tiguib

As in Calagcalag, the ARs and FADs in Tiguib were destroyed or deteriorated, but in contrast to the first case never reinstalled. The fish sanctuary was not installed at all. The mangrove reforestation survival rate was only 10 per cent and livestock dispersal was limited. Apparently, sea-farming was not part of the project. The Fishermen's Association has had its 'ups and downs', but at the moment of the evaluation was active – the interest of the members mainly in mangrove reforestation and access rights to ARs and FADs, despite the fact that these were reported as destroyed at the time. The Fishermen's Association was

reactivated as a means of obtaining a milkfish fry collection concession. In this village there was a clear difference in the understanding of
the use of creating rules, ARs, and FADs, between the members of the
Fishermen's Association and the non-members.

The Tiguib nearshore component of CVRP-1 was considered to be less
than successful by project staff. From the perspective of the fisher
households interviewed, however, statistically significant improvements
were perceived in seven of ten impact indicators. The three remaining
impacts were perceived as improving, but were not statistically significant. Again, it is important to note that this overall perception of positive change exists despite partial or complete failure of some project
interventions. Fisher people noted an increased level of knowledge of
project objectives and a high level of participation in project planning
and changes in the project. They may have felt a sense of empowerment because while the original association failed, a new association
was formed to take advantage of an economic opportunity. This indicates that when the right circumstances existed, the residents of Tiguib
demonstrated that they could work together.

From the study it appeared that project staff were focusing more on
easily observable impacts, for example, functioning fishermen's organisations, area of mangrove successfully replanted, etc. The fishermen,
however, felt a sense of empowerment and increased knowledge. They
had more information with which to make decisions and improve their
life, they had more skills, and they felt more integrated into the economic and political mainstream. However, although the general project approach was bottom-up, indicators for project success were solely
defined by the project implementers or outside evaluators.

The bottom-up approach implies a highly participatory form of planning, action and evaluation, that is, one which involves a high degree of
input by the local participants (IUCN, UNEP, WWF 1991). Hence, evaluations by *both* project staff and beneficiaries are important, and it is also
important to understand that they reveal different results based on different criteria of success or failure. In fact, it is the evaluation by the community members themselves that will influence their subsequent
behaviour and thus the potential sustainability *and* success of the project.

3. From development to people-centred resource management

Evaluation of intermediate objectives is basic to any project evaluation.
It is a matter of determining stated objectives from project planning

documents, and determining whether or not the objectives have been met (Pomeroy *et al*. 1996). It is clear from the above, however, that this way of assessing project success is inadequate, but then, *why* are such indicators still used? In order to understand this, we have to review the historical practice of development project implementation.

Indicators commonly used to measure success of development projects were designed at a time when most projects aimed to increase the well-being of the local population by intensifying their (agricultural) production. Local people were merely recipients of advice and training to increase their harvests. The top-down, blueprint development approach stimulated neither people's affinity with nor their sense of responsibility over the projects. Failure of many of these projects in the 1960s and 1970s led to a shift in development thinking, of which Chambers' *Farmers First* (1989) is an enlightened example. Attention shifted from strictly production to farming systems research and extension. It was acknowledged that without the support, consent and participation of the target population, a project was likely to fail (Korten 1986). Technology and production, however, were still regarded as keys to development.

The ever-increasing exploitation of natural resources, however, led to their degradation and over-exploitation. In the early 1980s, it became clear that sustainable resource use and conservation were the only options to reverse the process of decline (Rio Summit; *Our Common Future* 1987; *Caring for the Earth*, IUCN 1991). The efforts of centrally organised conservation projects, such as massive reforestation projects, to combat over-exploitation and degradation and to improve monitoring, control and surveillance, however, also proved ineffective (see also Persoon and Sajise 1997). Governments were often unable to implement or enforce the regulations that were issued for the management of natural resources. At the same time, local communities did not have the means and mandate to participate, even if they had a traditional management system. Slowly the idea emerged that it was not the government institutions, but the people who were the true resource managers (Korten 1986; Poffenberger 1990a; Bromley 1992).

Community-based Resource Management (CBRM) strives for more active people's participation in the planning and implementation of natural resources management. Its central concern is the empowerment of groups and social actors and to create a sense of self-reliance. It starts from the basic premise that people have the innate capacity to understand and act on their own problems (Ferrer and Nozawa 1997). However, it has also been realised that the government is a crucial partner in resource management to provide enabling legislation, information and other assistance (Pomeroy and Berkes 1997). This form of management is called co-management, a partnership between government and community.

A wealth of literature has been written on Community-based Resource Management and Co-management (for instance, Jentoft 1989; Pinkerton 1989; Berkes 1990; McCay and Acheson 1990; Ostrom 1990; Poffenberger 1990; Bromley 1992; Pomeroy 1994, 1998). Many lessons have been learned conditions for success defined, but *how* to measure this success, or rather, the true perception of success, is still unclear. Although the approach has changed over the years, and more components are included to enhance project success, the indicators used to measure this success have not been changed.

4. Shortcomings in evaluation

The question therefore is: what kind of indicators do reflect project success? For the answer, we have to look at the structure of CBRM and co-management itself. In order to shift from the role of resource users to the role of resource managers, project participants in CBRM required different skills and capabilities. A new set of methods were designed, called participatory methods, through which the local population would not only have a voice, but would also be able to participate in all project components.

A few examples currently in use are: PRA (Participatory Rural Appraisal), which focuses on interaction based on communication means which are locally understandable PTD (Participatory Technology Development) a method which includes local production techniques in agricultural development; and RAAKS (Rapid Appraisal of Agricultural Knowledge Systems), a method which is used to learn about a local situation and, through dynamic interaction, works at problem resolution, in, for example, agricultural settings (Kidd 1979; Korten 1986; Uphoff 1986, 1992; Engel *et al.* 1994). In addition to the use of participatory research methods, in many projects, a training component is added to increase local capacity. People need to acquire communication skills, the ability to formulate project goals, to plan and delegate tasks, solve problems and other abilities necessary to be partners in executing a community-based management project, a process usually referred to as capacity-building.

If we take a look at the general process of implementing a CBRM project, various phases can be distinguished. In his article on CBRM,

Pomeroy (1998) distinguishes three project phases: the pre-implementation phase, the implementation phase and the post-implementation (or phase-out) phase. The first phase (sometimes overlapping with the second) is usually the phase in which the community members develop their 'management skills'. This phase of community development is also referred to as the 'social preparation process' (Pomeroy 1998).

The Philippine experience shows, as do other cases, that the social preparation process is time-consuming. It was concluded that for a truly participatory project, the time required for people to master new skills for CBRM would be at least three to five years, but probably longer (Borrini-Feyerabend 1996; Pomeroy *et al.* 1996; Mulekom and Tria 1997; Mulekom 1998). In reality, however, many projects are planned for a shorter time span, even if they do include a social preparation process! The result is that the material project interventions are carried out while the beneficiaries are still in the process of developing the skills needed to understand the project interventions, and are not in a position to implement them. Consequently, at the end of the project life – which is not necessarily the end of the developments in the field – project interventions are not (or not fully) implemented or sustained. In these cases, the project is deemed to be a failure. However, it is entirely possible that if the participants had been given more time, chances are that in the longer run, project objectives may have been accomplished.

Not only is the timing to measure project success essential, it is also important to know *what* to measure. The fact is that even though community-based resource management is people-centred, in most cases project evaluation does not include the personal achievements of the participants. Despite the personal development of the project beneficiaries, project evaluation has remained focused on quantitative and material indicators of project success. The less tangible results of the project such as changes in attitudes, beliefs and values of the project participants, or their sense of empowerment, which was important in the case-studies mentioned earlier, were and are not measured.

From the above, it can be concluded that the reason why projects are often evaluated as being unsuccessful is because neither the timing of the evaluation, nor the criteria used to measure success is appropriate. Chances are that the criteria which are measured are those which the project implementers feel are important, while overlooking those representing the experience of the participants. The obvious solution to this deficiency is to redefine success and to develop new criteria to measure project results. This is, however, only possible if those who evaluate the projects are aware of their limited perception of reality, and recognise that the local community may evaluate project success differently to the researcher or evaluator him- or herself.

5. Emic and etic

The disparity between what is actually measured and the people's perceived impacts of the project, can perhaps be understood using some theoretical ideas from anthropology. Harris (1991: 20), for example, writes:

> The problem is that both the thoughts and behaviour of the project participants can be viewed from two perspectives: from that of the participants themselves and that of the observers. In both instances, scientific, objective accounts of the mental and behavioural fields are possible. In the first instance, the observers employ concepts and distinctions that are meaningful and appropriate to the participants; in the second instance they employ concepts that are meaningful and appropriate to the observers. The first way of studying culture [or perceptions of success] is called *emics* and the second way is called *etics*.

Borrowed from linguistics (Pike 1954), phon*emic* refers to what a sound signifies in the minds of the users. Phon*etic*, on the other hand, refers to scientific descriptions of sound with no reference to meaning, that is, from the outside. Etic categories are those that the researcher employs for the purposes of scientific classification, analysis and understanding of human–environmental interactions (Lovelace 1984). Emic is concerned with the elements, aspects and interpretations of the belief system as perceived or conceived by the members of the culture or society under consideration (see also de Groot 1992 on the use of emic and etic in the perception of environmental problems).

The problem is that even though many projects claim to value the participation of the local people, they tend to neglect emic categories. Project output is expressed in concrete, technical terms; evaluations focus exclusively on etic observations. However, strictly etic assessments are inadequate for projects that have socio-cultural impacts. 'During the process phase it is meaningful to determine how the people perceive the natural environment, the local problems, the alternative solutions, their abilities to intervene, but most importantly, their capabilities to do this collectively,' write Pomeroy and Carlos (1996). In the absence of emic considerations, it is impossible to discover these local conceptions and perceptions.

This does not mean that etic measurements cannot be used to measure success. On the contrary, the physical aspects (that form part of the etic environment connected to the values and views of the project implementers) are important project results. However, they need to be measured in a later phase. It is essential that the two types of project evaluation take place at the appropriate time. As long as the social preparation phase is ongoing, there is no need (or use) to measure material output even though they are part of project activities. This is the moment for emic considerations. Only after the project has phased out and when the community has had the time to carry out project interventions, is it appropriate to measure according to etic standards.

5.1   Subjectivity objectified; or, emic becomes etic

The purpose of the discussion of emic and etic is not only to elucidate the difference in perspective, it also touches on issues such as objectivity and validity. What we have seen is that not only the timing, but also the decision of *what* to measure is crucial in the evaluation of project success. Until a more pluralistic approach is widely accepted, what is measured depends for a large part on the demands and requirements of donors and/or the implementing agency (see also Anderson *et al.* 1998). The prevailing positivist mind-frame of researchers – who often design or are involved in implementation of the projects – evokes a natural preference for easily observable results that can be quantified and measured (Leurs 1996). However, it is a false assumption that only empirical observations lead to valid and objective measurements. The non-material, subjective, personal experience of the project participants can also be transformed into observable facts.

De Groot (1992) explains in his environmental science theory how people form conscious structured pictures in their heads based on facts from the outside world. He concludes that if we externalise this subjective picture – by putting it onto paper, measuring it and testing it – it becomes objective. During the social preparation process of a project, people not only learn, but become conscious of their personal progress and capacities. This personal development needs to be translated into 'pictures'. With the right set of indicators and methodology people's perceptions can be captured; hence, their experience can be externalised and measured. This is exactly what is foreseen in an emic evaluation: the subjective, emic experience becomes an objective, etic observation that can be measured and analysed in a scientifically valid way.

There is another advantage to this approach. During the social preparation process, it is expected that people's perceptions will change and that their awareness concerning environmental problems will grow. People will realise that not only do they cause environmental problems, but they also can play an active role in solving them. At the moment when this consciousness becomes part of their conceptual framework, we speak of *internalisation*: the process in which behavioural codes originating from others become part of the personality of an individual (Wilterdink and van Heerikhuizen 1985). The measurement of people's perceptions and experience at an early stage stimulates the process of internalisation and the project beneficiaries become more self-aware. Instead of following passively what is prescribed by the project, people will start acting based on personal considerations.

Increased self-awareness further enables project participants to contribute and join the process as equal partners in sessions where project goals are developed with the project implementers. This is extremely important during the pre-implementation phase of co-management, because it is often in this phase that a resource problem is recognised by the resource users and stakeholders, followed by an open discussion about the problem, negotiation, consensus-building and the development of agreement on a plan of action and strategy (Pomeroy 1998). At this stage, the community may seek assistance from outside agencies such as government, NGOs and others.

In the next phase, the implementation phase, fieldworkers and community organisers generally conduct meetings to discuss the concept of co-management and discuss the project. Baseline data are collected and participatory research conducted. Once the community feels comfortable with the community organisers, a community problem, needs and opportunity assessment is conducted, information is shared, and management and development objectives are defined in a communal process (Pomeroy 1998). This is, of course, in an ideal situation where there is consensus and where there is common ground to define the project objectives. If this is the case and the outcomes are the result of a communal activity, it may be assumed that that the physical objectives of the project as perceived by the beneficiaries (emic) and the implementers (etic) coincide. The second measurement of these physical outcomes, right after the project phases out, thus will represent an objective, etic perception of project success.

6. What to measure and when?

Two important components of the social preparation phase are 'communication-mechanisms' and participation. Communication mechanisms are used to clarify and define the roles of participants, that is, local people and project representatives, in the different phases of the project.

Exercises can help people to understand principles of adult learning and community participation. Trainers and participants become aware of preconceptions about each other; they learn about group behaviour and role perceptions (see for example the FAO Handbook for Participatory Evaluation 1988; UNDP Toolbook 1993). These exercises and other tools enhance participation. They help to establish a working climate that stimulates involvement of various stakeholder groups and allows people to partake in decision-making. The communication mechanisms thus provide the prior conditions for active participation and stimulates capacity building.

From several studies and experiences, one criticism of participatory methods is that, despite the attempts to develop a methodology which creates equity, PRA and other participatory methods still imply an imbalance of power (possibly because of funding) and control (due to lack of accountability) (Leurs 1996). The construction of local knowledge (and the emic picture) is strongly influenced by existing social relationships, in particular by relations of gender, power and by the PRA facilitators themselves (ibid.). This means that these methods need to be applied critically.

Several studies, guidelines and training manuals for a participatory approach provide possible indicators that represent community attributes, such as awareness and capability (UNDP Toolbook 1993; Pretty 1994; Pomeroy *et al.* 1996; Borrini-Feyerabend 1997; IIRR 1998). These indicators allow a measurement of project success on different levels: the personal/individual level and the community level. It is recognised that these indicators are predefined and, in this sense, contradictory to what we actually want, that is, development of indicators in the field. However, in the absence of a more adequate method, this seems a good basis for the measurement of project results. It is hoped the discussion in this chapter and the growing recognition of the need for a more pluralistic approach (Anderson *et al.* 1998) will stimulate the development of new tools for evaluation.

Possible indicators that represent project beneficiaries' personal achievements and benefits are listed below.

Individual indicators for project success

- *Involvement*: in the project design; in decision-making; in management; in defining boundaries; in rule development.
- *Capability*: to express an opinion; to make decisions; to prioritise issues; to participate in a meeting; to write a proposal; to speak in public; to work in committees.
- *Control*: over the process; over resources; over people's own life.

- *Access*: to knowledge; to meetings; to resources.
- *Skills*: to repair and maintain technical equipment; to manage a project; to solve problems.
- *Personal change*: in awareness; in sense of responsibility; in self confidence; in initiative; in self respect; in generating new ideas; in willingness to deviate from customs and community values; in willingness to take risks.

With these indicators, differences within the community can be measured (in terms of class and gender, for example). For the community as a whole, success will be defined in a different way because achievements other than the purely personal are also important. The community viewpoint is naturally more concerned with general benefits and accomplishments. A set of indicators to identify changes on the community level is listed below.

Community indicators for project success

- *Communication*: commitment of stakeholders; recognition of stakeholders; understanding between stakeholder groups; expression of different viewpoints; level of open disagreement.
- *Representation*: of various stakeholders; of various social groups; of women; of socially marginalised groups.
- *Collaboration*: between individuals; between neighbourhood groups; between various social (differentiated) groups.
- *Trust*: between staff members; between staff and government; between staff and project beneficiaries.
- *Support*: of higher government levels; of the local leaders; of a NGO; of the project staff; of village-based organisations.

It should be apparent that this list is not exhaustive. Furthermore, it is likely that selection of indicators will be useful depending on the local situation. However, with such indicators emic evaluation can be attempted. The early emic evaluation has the advantage that it allows the project to adapt over time and thus prevent possible failures (see also Pollnac 1989). Since the social preparation process often takes place during the implementation phase when physical project activities have also started, it may be an option to include material and physical project indicators in the preliminary evaluation (see list below). This evaluation typically focuses on the process of project development in terms of material achievements. It will show whether the project goals are appropriate, need to be modified, adapted or skipped altogether. The advantage of this early measurement of physical indicators is that it enhances the self-esteem and awareness of participants, and also reveals the priorities of the project participants. Hence, it further stimulates internalisation of project objectives, which makes it more likely that the selected activities are actually carried out.

It is only in the last phase of the project, the post-implementation phase, that the final etic evaluation of the project takes place. Not only are the physical outputs of the project measured in a quantitative way, but also the organisational, non-material, success factors are quantified, such as the intensity of group involvement and the functioning of the management system and enforcement mechanism.

Project indicators of success

- *Success in terms of material output*: size of yields; catches per unit effort; hectares of protected areas; hectares of mangroves/forests replanted; occurrence of destructive practices by local people.
- *Success in terms of human involvement*: number of people attending the training; numbers of participants in project; frequency of staff-meetings; size of the network.
- *Success in terms of project benefits*: division of benefits; economic opportunities; well-being in terms of health; well-being in terms of income; flow of investments; education level.
- *Success in terms of management structure*: management institution designed and active; management plan and regulations designed and implemented; enforcement structure in place; conflict solving mechanism in place; leadership.
- *Success in terms of participation*: type of participation; dimension of participation.

Ideally, the indicators would be selected by the project participants during the implementation phase of the project. In reality, however, many indicators are predefined in a project proposal drafted by the implementing or funding agency. Even though this may be the case, it is still advisable to lead all project participants through a process in which the goals and objectives are discussed and prioritised. This is essential because if the project goals are locally derived and internalised and not *imposed on* the people, the definition of success for both the implementers and the participants can be expected to be similar. This final evaluation, then, will truly represent project success.

7. Methodology

The last question to be answered is: *how* should project success be measured and by *whom*? The emic and etic measurements require a standard methodology which is valid and allows comparison. The core of project evaluation is people's perceptions. It is essential to use the right method to measure these perceptions, especially since for a number of these indicators no base-line data is available. The measurement of people's perceptions is complex. Perceptions cannot simply be measured by asking people 'what they think', as happens in many studies. These kinds of questions do not reflect the complexity of people's thoughts and the subconscious. Emic indicators (such as perceptions and attitudes) are non-material and qualitative yet quantifiable and demand a certain approach in order to be measured (Bernard 1994).

There are various ways to measure project success and to operationalise the indicators. The methodology to measure people's perceptions of success depends largely on whether the evaluation is action-oriented or is part of an academic exercise. In the latter case, each indicator needs to be thoroughly operationalised and studied. This could be done, for example, through anthropological fieldwork directed at the development of a set of indicators. This would lead to scientifically valid outcomes, but can be time-consuming, especially when it concerns a long list of indicators.

Where time is limited because action is required, or where funds and the availability of skilled researchers are limited, alternative methods need to be used. One example of such a method is a visual self-anchoring ladder scale used in the ICLARM-IFM Fisheries Co-management Project (Pomeroy *et al.* 1996). This base-line independent method allows for the use of non-parametric statistical techniques and makes use of the human ability to make graded ordinal judgements. Fishermen are asked to answer questions about the state of the resource, fish catches, personal well-being, income, occurrence of conflicts, collective action etc. by using a picture of a ladder with ten rungs as a visual aid. The lowest rung represents the worst possible condition, the highest rung represents the best. In the study, fisher people's perceptions were recorded of past conditions, current conditions and degree of optimism for the future. The technique deals with variability in perceptions over time and facilitates analysis of the perceived project impacts.

This is only one example to show the use of participatory techniques to measure project impacts. Other methods, such as participatory evaluation methods, may also be useful (FAO 1988). These techniques allow a great deal of input from participants and are very useful for rapid assessment of the local situation (Chambers 1991; Drijver 1993; Mosse 1994; Blauert and Quintanar 1997). The exercises can be adapted to measure personal change and development accruing from the project, and outcomes can be

1. **Visual scoring and ranking systems** can be used to measure changes in wealth and well-being, development of skills, representation of social groups, etc.
2. **Time lines** can be used to represent significant changes in the village, but also on the individual level.
3. **Seasonal patterns** can be used to show the relative magnitude of workload, they can also illustrate project activities and extent of involvement in the project.
4. **Venn and linkage diagrams** are useful to represent social relationships or the importance and influence of different individuals or institutions.
5. **Visual estimations and quantification** record such things as yields and prices, but can also be used to measure skills, initiative, commitment, etc.

Figure 8.1 PRA techniques

Sources: Adapted from Jiggins and de Zeeuw, 1992; Pido *et al.* 1996.

quantified and compared (see Figure 8.1). For more easily quantifiable indicators of project success, relatively straightforward methods could be used, such as observations, enumeration (census) and surveys.

8. Conclusion

Over the last decades, development projects have shifted their approach from development to a people-oriented approach. An important concern of CBRM and co-management is the empowerment of groups and social actors. These approaches require extensive participation and the development of local capacity. Project participants need to develop the skills required to manage their resources. However, the personal development of project participants is often not evaluated and project evaluation remains exclusively focused on material outputs.

Project success depends largely on what is actually measured, when, and by whom. In order to evaluate project success from the perspective of both participants and implementers, we need to adapt the indicators used to evaluate the project. The personal development of the participants in terms of increased involvement, access, control, capability, skills and personal change can be reflected in an emic evaluation. These skills are largely acquired during the social preparation process and the appropriate moment to measure these non-material project impacts is directly after the implementation phase.

The early emic evaluation has the advantage that it allows the project to adapt strategies and adjust project goals, and thus prevent possible failures. It also provides a picture of the performance of the project over time, which may result in a more accurate assessment of what the project has achieved. In this way the chances of project success will not only increase, but it is also more likely that after the project terminates, the participants will continue the project's activities.

The fact that the material project goals are defined collectively with the assistance of government, NGOs and donor agencies, leads to the internalisation of these material project goals by all parties. Since the project outputs are agreed upon and based on collective decision making, it is more likely that the physical objectives of the project as perceived by the beneficiaries and the implementers will coincide and will be actually carried out. A second measurement at the post-implementation phase, focusing on the material project outputs, thus can be seen to be objective and would truly represent project success.

There are three critical points in this discussion: (1) the acknowledgement of a social preparation process, (2) the need to define project goals communally, and (3) intermediary measurements of various sets of project indicators. Current development structures, however, make no allowance for the extra set of indicators needed to evaluate project success at the emic level. Targets are usually set by those outside the community. Only in a few cases do the opinions of the participants play a major role in project design, implementation and evaluation. Furthermore, in many cases projects are carried over in too short a time, without a clear or long enough social preparation process, and with material interventions started too early. Hence, it is not surprising that the material interventions are either not sustained or never implemented at all. Consequently, the project fails to measure the non-material successes that may actually be experienced by the participants.

Without emic assessments, the evaluation of a project is not complete. But this is only possible when donors are prepared to change their approach away from predefined, entirely material project goals. The implications for donors are significant. It means a restructuring of project proposals to include a longer preparatory process, a redefinition of project goals, and possibly a longer implementation period. This has financial consequences, but more importantly, changing the approach would imply a drastic shift in authority and control over the project. Numerous evaluations and studies of failed projects are a clear indicator that these changes are required to increase the likelihood of project success in the short term, and thus to ensure sustainable resource use over the longer term.

9
Joint Forest Management: A Silent Revolution among Forest Department Staff?

Roger Jeffery, Nandini Sundar and Pradeep Khanna

1. Introduction[1]

> We have to remember that the forest department has been working over the years in a certain way and if we ask them to change overnight, resistance is bound to come. (G/S/1)

> The most important transformation that Joint Forest Management has brought about is the transparency in the working of the forest department. (AP/S/3)

> The change in the forest department is visible, as they have come out of their seclusion, out of their shell, their basic philosophy has changed. It has lead to change in decision making pattern due to interaction with the villagers. (O/S/3)

In this chapter we examine the discourses of participation in Forest Departments in four Indian states, and discuss the extent to which the views expressed in these quotes represent the changes taking place under the spur of newly introduced participatory approaches to forest management. After setting the context of debates about popular participation in natural resource management, we briefly discuss the research on which this chapter is based and contextualise India's introduction of Joint Forest Management, or JFM. We then consider the evidence on the general values held by forest staff and their specific views on JFM. Because almost all FD staff claim (in public) to support the extension of participation, we have looked more closely at expressions and perceptions of discontent, in all cases contrasting views at different levels of the FD hierarchy. We conclude that our evidence suggests that there is a good chance that outright opposition to participatory management will remain muted, but that tangible evidence for whole-hearted support is still insufficient to conclude that the programme will be a success.

By comparison with the academic analysis of the characteristics of the 'community' that facilitate success in participatory approaches for natural resource management in developing countries (Baland and Platteau 1996), much less attention has been paid to the natural resource departments of governments, and the changes in their bureaucracies that are necessary for participatory approaches to be introduced and supported. In India, supporters of JFM have assumed that the State Forest Department (FD) would be willing to manage parts of the forest jointly with community representatives, involving non-governmental organisations (NGOs) where appropriate, and have suggested ways in which this process might be encouraged to come about (Karnataka Forest Department 1994–96; Maheshwari and Moosvi n.d.). But empirical studies of the effects of the introduction of participatory approaches on forest departments are much less common (perhaps because many of the written accounts have been funded by FDs themselves).

Empowering local groups requires agencies to give up some of their authority, which 'demands a strong political commitment to the devolution of power on the part of the bureaucracy' (Poffenberger 1990: 102). We can identify two contradictory views on the likelihood of this happening. There is a 'widely held notion that most state agencies are centralized, authoritarian, formalistic, inefficient bureaucracies incapable of experimentation, self-critical learning or imaginative change' (Thompson 1995: 1521). Critics of the Indian FDs assume that they will therefore never become fully supportive of participatory approaches. On the other hand, some authors (particularly from within the FDs themselves) point to examples of change (such as the introduction of Social Forestry in the 1970s). A criticism of FDs (their hierarchical structures) is sometimes presented as a strength: once the orders in support of a participatory programme are issued in the State capital, lower-level bureaucrats may have to obey.

Some of the many problems that restrict the possibilities of major changes within FDs cannot be resolved within the FD itself. In common with the public sector in many developing countries, such problems include 'low salary levels in the public service as a whole, lack of effective performance standards, inability to fire people, too few rewards for good performance, recruitment procedures that did not attract appropriately trained people, and promotion patterns based too much on seniority or patronage and too little on performance' (Grindle and Hildebrand 1995). None the less, we argue that policy change towards more participatory approaches within the FDs of Indian states is possible, even without major structural changes at the level of society at large. Decision-makers not only have some 'space for defining the content, timing and sequencing of reform initiatives' (Grindle and Thomas 1991: 19), but also 'field-level bureaucrats are not just linear extensions of a hierarchical chain of command, and ... their acceptance of the participatory agenda is an important determinant of its potential success' (Vira 1999: 256). Those charged (often by donor agencies) with introducing administrative change also favour this middle-ground position: that FDs can become flexible, with an 'ushering of transparency into the department's transactions with most stakeholders' (Maheshwari and Moosvi n.d.: 5) and move from a 'culture of control to a culture of commitment' (ibid.: 25) through processes of institutional development. The apparently strong hierarchical structure of the FD also masks considerable areas of freedom for some middle managers to take innovative action (Vira 1999: 267).

This chapter assesses some evidence on attitudes and practices of FD staff from our recent study of the introduction of JFM in four Indian states: Andhra Pradesh (AP), Gujarat (G), Madhya Pradesh (MP) and Orissa (O). For simplicity of presentation we have grouped staff into three broad categories: field-level (Forest or Beat Guards, Range Assistants, Forest Rangers or Range Officers); middle managers (Assistant Conservators of Forests [ACF] and Divisional Forest Officers [DFO] or Deputy Conservators of Forests [DCF]); and senior policy-makers (Conservators of Forests [CF], Chief Conservators [CCF] and the Principal Chief Conservator [PCCF]) in the State capitals.[2] Three different kinds of interviews were carried out: structured questionnaires given to middle managers and field-level staff; semi-structured interviews with senior staff; and semi- and unstructured interviews with staff of all grades. (For details of FD structures in India, see Bahuguna and Luthra 1991; Saxena n.d.) This interview material is complemented with observational material collected during fieldwork over a 27-month period from April 1995 to July 1997.

2. JFM in India

During the 1980s, a generalised perception of the increasing loss of forest cover on land owned and managed by the FD was reinforced by the first results of satellite imagery. The Social Forestry Programme, dating from 1976 and involving tree planting on private and community land, seemed to be failing. In 1988 a new Forestry Policy was announced

(Ministry of Environment and Forests 1988). This was followed in 1990 by a circular letter, issued by the Government of India's Ministry of Environment and Forests, which encouraged the state governments to implement programmes of forest protection on degraded land in partnership with committees of forest users, with the involvement of NGOs where possible (Ministry of Environment and Forests 1990; see Yadama and DeWeese Boyd, this volume, for further details and background to India's JFM programme). Of the states included in our study, Gujarat and Orissa began implementing JFM soon after 1988; AP followed after 1992; and (apart from some well-documented experimental programmes) MP did not begin a full programme of JFM until 1995.

In general, four reasons explain why public sector agencies might introduce participatory approaches (Thompson 1995: 1521–2): a fiscal crisis, enhanced by economic liberalisation; donor pressures in favour of accountability and transparency; recognition of the failures of past approaches; and the demonstration effects of successful participatory programmes by 'third sector institutions' and other state agencies. All four apply, in different measures, to the Indian forestry sector. All four states in our study are facing pressures to reduce the size of their budgets and their bureaucracies, though these pressures do not loom large even in the perspectives of senior forest officials. There is, for example, little evidence that FD staffing complements have been reduced in the past ten years. In two of the states, AP and MP, JFM has been introduced with the backing of major World Bank-funded Forest Projects, but in Orissa and Gujarat donors have not been involved. In AP, the FD was divided: considerable interest in JFM prior to the negotiations over the World Bank project coexisted with opposition to any major change. Now, while some staff see the introduction of JFM as a homegrown idea, others see pressure from the World Bank as an important catalyst.[3] In all four states, resources from the World Food Programme have sometimes been channelled through the FD and used to provide benefits to communities and individuals as incentives for them to be involved in participatory initiatives. FD staff in all four states accept that the previous models of forest management had failed to protect good forest or to rehabilitate degraded forest land. Alternative models have encouraged change: in Gujarat some of the innovative NGO-funded projects (like the Aga Khan Rural Support Programme) were active in the forest sector in the 1980s. In Orissa, long-established community forest management arrangements have historically had little or no FD involvement (Sundar *et al.* 1996). The successful innovations and experimental programmes associated with Arabari in West Bengal are cited as spurs to action by some officials who visited the project as part of orientation tours before JFM was introduced in their states.[4] As one put it:

> I went to West Bengal as a team leader of a delegation to see the JFM in 1989. Before that I did not believe in it and felt that those talking about it must be aliens. But after a three day visit to Arabari, I was a changed man. (AP/S/1)

A continuum of attitudes towards community participation can be identified (Vira 1999, derived from Midgley 1986). A participatory mode is fully supportive of community involvement; an incremental mode is one with official but ineffective support, leading to ambivalence; a manipulative mode is one where support is designed solely to meet ulterior motives of the state; and an anti-participatory mode is uninterested or hostile to community involvement. We do not see these as mutually exclusive: different individuals within a state may have different views, and may change their views rapidly in response to rewards and penalties, or the responses of other actors. They may also hold several different positions simultaneously, concerning different aspects of the initiative. Here we ask how far the staff charged with implementing the participatory approaches support them, and for what reasons. How do they see the current and future emphases in their work? How far do current social structures support or discourage changes in attitudes and practice towards a more participatory mode? What is their experience of JFM so far, and what predictions can be made about the medium-term chances of success in achieving its goals?

3. Values and goals of Forest Department staff

Our respondents share many of the critical perceptions of the FD often reported in the literature.[5] The weaknesses of the FD are described as shortages of finance, with inadequately trained staff, unrealistic and unclear goals, with inadequate or disjointed strategies to achieve them. The FD is also described as hierarchical, though this is not always seen as a weakness. Paradoxically, perhaps, they describe the strengths of the FD as its discipline, the technical knowledge held by its staff, and its ability to work in adverse circumstances, to achieve its goals in remote areas. In many cases, staff report with pride that FD staff are the only government employees who regularly visit villages far from the main centres of population. In this context, statements of commitment to the re-orientation of the FD's management and orientation are not inevitable. How far do FD staff see the new approaches as contributing to their own job satisfaction? Some hints come from their answers to the questions on the reasons for which they chose the FD for a career, their view of whether it was a correct choice, and their expressed preferences for different kinds of postings within the FD.

As in the rest of the public sector in India, reasons for joining the forest service were varied. Most of those in the field-level (Range Foresters and below) and middle-management positions (ACF and DFO) talked in terms of applying for a wide range of positions, and joining the FD just because they were selected and offered a post. Thus some Foresters had joined the defence forces, the electricity board, a bank, the police, and the Education Department, but resigned because of better prospects in forestry, the chance of permanent employment or family opposition to alternative posts.

Only a small proportion talked of a long-term interest in forestry, and this was often mixed with the reality of what job was available. One case of a more determined commitment to working in forests was an ACF, who claimed a 'strong affinity towards plants'; one Range Assistant also talked of being a villager and wanting a job that was close to nature. Despite the requirement for the senior staff to have a scientific training, then, a concern for a 'scientific' approach to forestry was not central to career choices, and many staff described other motivations, including a wish to serve. Whatever it may have been in the past (Grove 1998; Rajan 1998), the Forest Service cannot now be characterised simply as one dominated by a scientific ideology. Similarly, opposition to change cannot be merely put down to the effect of outlooks and practices inculcated in school, college or in post-entry training.

Given these ambivalent reasons for joining the FD, it is not surprising (but perhaps somewhat alarming) that almost two-thirds of middle managers would not choose a career in the FD if they started out again. But our research provides no evidence that these managers do not want to continue in the FD because of the recent shift towards more 'person-oriented' rather than technical work. In all four states, whether or not they came to forestry from an interest in science, foresters accept that people need to be much more involved in forest management. These staff are concerned more with issues of promotion, status and recognition, as can be seen in their perceptions of good postings, self-assessments of the successes and failures of their careers, and their perceptions of the desired future ideal form of the FD.

The common perception of a good posting is the section concerned with managing the commercially valuable forests, though for a variety of reasons. For some, it is the discretionary power, resources and authority that come from posts there; as one Forester described it, 'thousands of people bow to them'. For others it is the challenge that the posts provide: the 'scope to do work'. Others desirable postings are in the Forest Development Corporations or in special divisions dealing with *kendu* leaf which offer 'money' because they involve awarding contracts.

We gave staff an opportunity to assess the strengths and weaknesses of the FD for a career. The levels of dissatisfaction seem similar to those expressed by other public sector employees. Middle managers expressed considerable dissatisfaction with their own careers. In some states the major complaints were the poor infrastructural facilities and living conditions available to staff, often expected to live in isolated areas, something that might be acceptable in the early stages but was hard to accept with children of school age. More generally, key aspects of the management structure of the FDs (the confidential record system, and the punishment and transfer policies), which exemplify for many people the hierarchical and authoritarian aspects of the FD, were picked out for negative comment. Middle managerial staff seemed relatively satisfied with their own achievements: what upset them was a perceived lack of promotion opportunities and a feeling that senior staff did not recognise their achievements. They suggested that the management system rewarded risk-averse career strategies rather than initiative and achievement; that promotion was almost entirely via length of service; and that the system of confidential annual reports by superiors encouraged an authoritarian hierarchical management style.

Not surprisingly, there is a general acceptance by middle managers of the need for change within the FD. Furthermore, the main areas of desired change (apart from improvements to the pay, conditions, and promotion prospects for staff) were forms of working which would allow for better relationships between FD staff and forest dwellers. JFM allows this reorientation most clearly, and many of their comments were cast in terms of views on how this was being introduced and implemented.

4. Views on Joint Forest Management

A major difficulty in understanding how FDs are responding to the new initiatives is that almost all forest officials support the participatory rhetoric in public (Thompson 1995). In specifying their reasons, the views of senior staff, who are charged with implementing the new policies but have not had to revise their everyday work patterns, are very different from those of middle-level and field staff, who have. But this does not mean that senior and middle managers foresee no problems with JFM. Both groups understand that JFM must lead to changes in the organisation of the FD; that JFM is generating conflict within and between villages; and many of them are sceptical about the roles expected of NGOs. Some middle managers and field staff are also positively inclined towards participatory approaches in general, and JFM in particular, because they expect that villagers will not in practice play a significant role in planning and managing forest resources. We will address this latter issue with data from our observations of how JFM has actually been introduced.

4.1 Formal acceptance of the participatory rhetoric

Criticism of the participatory programmes is muted at best: officially, all are in favour of JFM. Thus the interviews with senior staff are punctuated by comments about the need for JFM as a solution to a downward spiral in forest quality, and that the involvement of villagers is a key to the long-term maintenance of the areas which have become degraded. Senior staff foresee 'a bright future' for JFM as a sustainable approach (G/S/1); that 'There is no alternative to people's participation. Even the villagers have understood it. Only with their help we can save the forest which is their lifeline' (MP/S/1). 'There is no alternative other than JFM, people have understood it and are not reluctant but very responsive. In JFM areas there is no manipulation by the Forest Department staff as there is transparency in dealings' (O/S/1).

Answers to questions about their current roles and the future ideal reveal that middle-level managerial staff (DFOs and ACFs) welcome a considerable expansion in their work interacting with forest dwellers, and liaising with other Departments to introduce rural development activities. They foresee much more time in the future being spent on protecting and monitoring the forests, rather than being administrators or grievance handlers. Their version of the current functions and the importance these functions should have in the future give further insights into the acceptance of a need for change. The middle managers expect that in the future revenue generation will be less important and the development of tribals and of educating forest dwellers to build awareness will become more significant. Middle managerial staff foresee, and seem to be accept, an increasing involvement of local people in planning and forest management. They also foresee a greater role for the private sector in degraded areas, the promotion of agro-forestry, and greater interaction with other departments and development agencies – all of which are entailed in the shift towards participatory approaches.[6] Field-level staff are perhaps the most enthusiastic about JFM: while acknowledging that it involves a loss of power, they are also often appreciative of the benefits it brings in terms of less stressful relationships with forest dwellers.[7]

4.2   Senior staff: perceptions of the acceptability of JFM to middle managers and field staff

FD staff, at all levels, portray themselves as much maligned but valiant defenders of the forest, functioning under severe limitations of person power and resources. They see the roots of JFM in the work of out-standing forest officials (in Arabari, for example). Some senior staff see JFM as a natural extension of FD activities only for the patches of land that the FD has been unable to protect adequately on its own. In their view, JFM has only limited implications for the way that the FD works on most of the land it owns. But other staff argue that considerable changes are needed in Departmental structure and administrative process, and that JFM is a vehicle for introducing such changes. Given the scale of transformation required, they see progress so far as very commendable, and perceive that junior staff have welcomed (or at least tolerated) the changes in their working practices:

> There has been a mixed reaction amongst the forest department staff towards JFM. Some are not convinced about it but their number is reducing gradually. It is a better way of working and we have to conduct ourselves in a different manner when dealing with the villagers. Our role has become a mixture of that of developmental workers and the police but an excess of any of the two will jeopardise the whole programme. JFM demands attitudinal changes. Staff should be trained repeatedly. Our training has been more oriented towards the higher staff than lower functionaries. We need to emphasise on training of lower staff also. (G/S/1)

Clearly, one difficulty faced by the FD staff is that in their previous operations they were using a very different style of interaction with forest dwellers:

> People were anti-forest department and the forest department also behaved as a para-military force. Some Forest Department staff misused their powers to coerce people for illegal gratification during

their visits to the villages, which gave Forest Department a bad name. This made the people anti-forest department.... Under JFM we have found a way to limit unrestricted entry by self-restraint by the people. (AP/S/1)

The policing job of the Forest Department is glamorous, but it needs to be kept in the background in today's context and participatory methods need to be the approach. This is what Joint Forest Management wants to achieve. Joint Forest Management is being implemented by the officers because they have been told to without any preconceived ideas. Disparity in the working style of the officers is pronounced and it is an uphill task to bring them to a common platform. Some people are corrupt and this is also a disadvantage as the people are finding it difficult to believe the sudden change in the Forest Department. (MP/S/4)

In part, the attraction of JFM work is that it is a key area of innovation within the FD, and offers the opportunity for ambitious staff to make a name for themselves.

The staff want to remain in the Joint Forest Management areas. They are more confident as they know a group of villagers is with them to help them at the time of any difficulty. They are feeling that they have more chances to reflect their efficiencies and come to the lime-light. (MP/S/1)

Some senior staff recognise, however, that the 'real' policy-makers may not have made sufficient effort to ensure that middle management and FD field staff have a sense of 'ownership' of the new policies:

Unfortunately no specialists make decisions: it is just the IAS or politicians. Policy decisions are not participatory, and the FD has no sense of participation because it didn't make the decisions. (DFO, MP/April 1995)

5. Criticisms of aspects of JFM policies

Those who are less than wholehearted in their support for JFM are few. None the less, many staff make specific criticisms of JFM, in terms of the problems it raises for them or for villagers. The following are the main areas of criticism.

5.1   Inter- and intra-village conflict is being generated

Several senior staff noted that conflicts were being exacerbated, or would become more likely, both within and between villages.

> Conflicts are there due to vested interest of the local people who want to amass disproportionate wealth. People are understanding this. Conflicts may arise due to rich class of the village but as yet we have no feedback. But indifference is maintained by some in the villages. We are facing inter-village conflicts also due to our own fault of allotting wrong areas for protection. We are trying to rectify wherever the areas come under the purview of another village but are protected by another. (MP/S/1)

> Local politics is the main stumbling block, besides the inter-village, intra-village or village-and-forest department conflict. Inter-village and intra-village conflict has developed due to the activities of party politics at the state-level. Just to prove their points the forest becomes the casualty in the village. (O/S/2)

On the other hand, some argue that at the very least, conflicts between the FD and the villagers have reduced:

> Conflict between the forest department and the villagers have reduced. The inter- or intra-village conflict still has to come to the surface. (O/S/1)

5.2   Relationships between forest protection committees and *panchayats*

The current structure of autonomous forest protection committees at the village level is not in accordance with the national pressure to revitalise the *panchayat* (village-level council). Like many other government departments, the FD prefers to work with its own village committees and regards the *panchayat* as too politicised and potentially corrupt:

> There is no link between the *panchayat* and the village institution as the *panchayat* is a political entity while the village institution is village based and will remain the same even with the change of any party at the state-level. (O/S/1)

Others recognise that in the longer-term, it will be necessary to forge stronger relationships between JFM committees and *panchayats*:

> Politicisation in the later stages as in case of the Panchayati Raj may create problems for forest development. It would then be difficult to take back the things which are in their hands in case a conflict arises. (MP/S/1)

5.3 Likely failure of JFM to deliver the promised rewards to communities that protect forests

A major perceived problem for JFM is that it depends on the FD being able to marshal enough resources to meet the subsistence needs of those protecting forests. Unfortunately, several say, the resources, rules and arrangements that have been made so far are inadequate for this task:

> Today Joint Forest Management has taken off with great success, it is inspiring but if the sustainability needs are not addressed properly then it may fail and repercussions may be serious. Forests will be the worst sufferers in such a scenario. In the policy formulation a lot of things have been left unsaid and ambiguities remain in issues like the benefit sharing, distribution, selling, marketing and even funds procurement. If the income generating schemes which have been undertaken do not yield the desired results, then what? (MP/S/4)

> JFM is not going to be a success, as people are not going to get the benefit as they have envisaged. It would have been better if we had promoted Agro-Forestry which would yield benefits much faster. (MP/S/5)

A lack of clarity in the rules and arrangements still needs to be addressed. In Orissa, because of the perceived likelihood that the benefits being offered under JFM (limiting the community share to 50 per cent) may be less than the benefits offered under similar land under the Social Forestry Programme, political pressures on the FD have increased:

> The constant demand of everyone (politicians, activists, NGOs) is 100% benefits for the people without giving due consideration to the scientific management of the forest. I am apprehensive that if in the forest department we also stop believing in scientific management of the forest then how can the forest survive? (O/S/2)

Two other problems were raised by staff. First, some forests did not provide enough short-term benefits for villagers. FD staff made up these
shortfalls by drawing on other resources, either those under the control
of the FD itself (especially in AP and MP, where there are World Bank-
funded projects) or from those under the government's own development departments (Tribal Welfare, or Rural Development). But some
staff reason that this undermines the JFM programme. Some villagers
may be tempted by the development benefits (for example, work on
constructing check dams, access roads or cattle-proof trenches) rather
than the intrinsic benefits from protecting the forest. Once the development benefits dry up, the forest protection committees are likely to
become less active. Similar problems have bedevilled the World Bank
eco-development programme, for example (Baviskar 1999).

Furthermore, in the World Bank-assisted states, the FD will not be
able to grant the same level of support to all JFM villages. Villages
which have not been taken up in the early rounds fear that World
Bank money will run out, and this causes resentment:

> The JFM policy is somewhat wrong, because it involves us giving
> development inputs. People in other villages get jealous of all the
> development inputs. In the JFM villages we are friends but in the
> other villages, we remain the policemen. (MP, Range Officer, field-
> notes)

5.4   Attitudes towards the roles of NGOs

FD staff voice the most uncertainty, sometimes outright opposition, to
the proposals on the roles of NGOs, proposals that show a clear continuity with general discussions of the best way to introduce sustainable
natural resource management (Farrington *et al.* 1993). The Government
of India documents display the following understanding of the potential
roles of NGOs (Ministry of Environment and Forests 1988; Ministry of
Environment and Forests 1990). First, there is a need for ('committed')
intermediaries to bridge the gap and promote trust and understanding
between public administration and people's organisations. Further,
NGOs are expected to 'motivate and organise' the villagers (not FD staff)
and NGO roles are therefore envisaged at the local and pragmatic level,
rather than at higher levels affecting policy and structural relations.
Finally, there is a concern that NGOs may be hastily created in order to
take advantage of new opportunities offered by the policy, or that NGOs
without appropriate expertise may try to play a part.

Following these policy statements, each state has outlined the roles of NGOs, the FD and the village forest protection committees. The net effect is to be much broader and more ambitious in expectations of NGOs than is suggested in the original documents. NGOs are expected to act as mediators between the FD and villagers where mutual trust is lacking (see Yadama, this volume for an example) particularly at village level, but also to a lesser extent at higher levels, through regional networking. They may educate and train FD staff, other NGOs and village communities about the potential forms of development that might ensue from JFM agreements (e.g. by running awareness-raising and discussion workshops, or facilitating study visits to areas where participatory forest management or other relevant forms of participatory development have been practised.) They can provide skills in participatory planning and learning methods, conflict management, and community organisation for training sessions are organised by the FD; and they are also asked to prepare training manuals and so on. In some states they have been asked to undertake research and documentation of the programme, and to communicate their findings through newsletters.

Nevertheless, it remains unclear whether NGOs are *necessary* for JFM to proceed. None of the resolutions offer clear guidance on distinguishing different kinds of NGO and NGO roles, that would help Forest Departments understand potential pitfalls and advantages in forming new partnerships with NGOs. Only their roles in facilitation are recognised by the state (Potter 1998; Thin *et al.* 1998). On the one hand, then, government sets out a certain number of defined and limited roles for NGOs in JFM. On the other hand, the ideology and structure of NGOs and roles performed by NGOs able to be involved in JFM are very diverse, despite the surprising degree of convergence on notions of community management.

Field staff in the four states are more sceptical about the increased roles of NGOs than they are of any other aspects of the current plans for JFM. Senior staff share these ambivalent responses to the potential roles of NGOs. NGOs may be able to play a catalytic role, but they are not spread equally across the states. They also cannot be created specially for JFM, and where the terms of an agreement with donor agencies requires an NGO presence, they may have to be encouraged into forestry work from their activities in other spheres of rural development.

In MP, several senior FD staff claim that NGOs do not exist, despite the weaknesses in FD structures. In the absence of female FD staff, and of NGOs working on forest issues, and of women in particular, our own research colleague was asked to fill the role of 'Lady NGO' in some of the training sessions held in Dewas Division. In addition, in MP some NGOs see the World Bank project as very problematic, and there is considerable hostility between them and MP FD on this account.

In Orissa, some areas have well-developed NGOs and others have very few, and opinions on NGO involvement are mixed. 'Though the entry of NGOs has been late in Orissa they have contributed a lot. They should restrict their role to motivation only and not try their hands on management' (O/S/2). Some NGOs also believe that some legal powers should be devolved to village institutions, whereas FD staff believe they should retain these powers: 'NGOs are over-enthusiastic. They are claiming certain rights on behalf of the villagers but have calmly forgotten their responsibilities' (O/S/4).

In Gujarat, where large and successful NGOs with forestry interests do exist, foresters may feel threatened by the implied or actual challenges they pose to the authority of the FD, for several reasons. These NGOs are usually better funded than the FD, with more access to transport, computers and foreign travel. The professionals are often better educated, paid more and are more fluent in English and in the jargon of participation. In Gujarat, foresters may accept the useful roles of NGOs in training and motivation, but they are not necessarily keen to see NGOs continuing to play a large role: 'To my mind due to [the] presence of [the] NGO we built up an atmosphere where there is competition which may lead to conflict also. So, the role should be defined for the Forest Department and the NGO, as overlapping of works disturbs the spirit to work together' (G/S/3). Another said that the contribution of NGOs has been mainly to criticise, and they need to accept the continuing role for a state-wide forest policy with a major role for the FD.

In Andhra, senior FD officials have less experience of working with NGOs: as one put it, 'very few NGOs have conviction and expertise on JFM' (AP/S/1). Senior officials are unclear about the long-term role of NGOs. Some see the village committee as the real NGO: 'NGOs should work as a facilitator till the Van Samrakshana Samiti (VSS) itself takes up the role of an NGO' (AP/S/2). Others see the long-term role of NGOs as organisers of networks of Forest Protection Committees, as in Orissa.

6. Attitudes and practice in the introduction of JFM

Despite some specific reservations about aspects of the JFM policy, most staff support the policy. But will this support be maintained through the changes that might be necessary to implement the new approach fully in practice? Our observations from the field suggest that 'people's participation' rarely has a major impact on the everyday work practices of field-level and middle-management staff, for three main reasons: first, few staff are prepared to go beyond the call of duty to infuse new procedures with a different spirit from that which underlies traditional practice within the FD; second, departmental structures have not changed sufficiently to take account of the new emphases; and third, surrounding norms and values, for example of staff in comparator Departments or in the rest of Indian society, leave the participatory schemes of the FD exposed as out of the ordinary.

In each state we have observed some staff displaying exemplary commitment to making the new policy succeed. FD staff have, for example, raised funds to get bail for members of villages when a JFM agreement was being negotiated, and have placed themselves in personal danger between warring villages or in front of hostile villagers who were threatening FD staff. But these are no more than exemplary: one or two cases in a forest division are not enough to make a new policy replace the old one. Furthermore, (as noted above) police work continued, and many field staff felt conflicting pressures on how they should relate to villagers.

For those who wish to implement changes, the absence of supportive change within the FD itself is a major source of frustration. For most staff, JFM responsibilities – for selecting villages, conducting participatory rural appraisal (PRA), drawing up micro-plans, convening FPC meetings, writing minutes of meetings and implementing decisions in terms of delivering seedlings or other materials – are merely added on to their other tasks. 'In fact DFOs have no time for JFM – they have files to look at' (DFO, MP/April 1995). Time-consuming court cases and administrative work associated with personnel management can easily increase through JFM. Although JFM involves more visits, no extra travel allowances had been sanctioned: staff were pressured to meet the extra costs from their own pockets, or to find ways to use funds sanctioned for different purposes.

Workloads in patrolling the forests now being protected by Forest Protection Committees (FPCs) may decrease. Some middle managers and field staff said that they now slept more regularly at home in bed. But others noted that the threats of illegal cutting and smuggling may be transferred to other parts of the reserved forests, making little overall difference to the amount of patrolling that had to be done. In the presence of technological change amongst the smugglers (better guns, radios, cutting equipment and trucks) and the slow provision of similar equipment for the FD, JFM might make little difference to the security of FD staff.

Middle managers and field staff also note the lack of change in the existing FD bureaucratic styles. In all the states, these staff feel that they are expected to be able to change their working practices after just a few short training courses. They were mostly appreciative of these courses: but in the absence of supportive changes in managerial styles, they were aware of the contrasts between hierarchical and non-participative FD structures being used to implement participative strategies with respect to villagers. In practice, they sometimes felt unable to implement the new methods on a large scale. Thus participatory activities remained *pro forma*, following routines and bearing little relationship to the micro-plans that were developed in the office, usually in English, and often according to a format derived from a model created elsewhere in the state. Staff were often unable to make more than verbal promises about the sharing of final harvests, or to implement other changes requested by villagers, because of a lack of support from higher officers – who in turn felt constrained by the need to follow the rules. Furthermore, FD staff were aware that the staff of other Departments – Police, Revenue, Rural or Tribal Development, for example – were not under similar pressures to treat villagers as equals. This sense of relative deprivation emerged in a number of comments, particularly when they mentioned the 'unfairness' of external criticism of their efforts.

7. Conclusion

The current situation with respect to the support of participatory initiatives in these four FDs can best be described as ambivalence. Some staff at each level within each FD are supportive of participation in its own right, and leadership (from politicians, administrators, senior FD professionals and middle management) is sufficient for the introduction of participatory approaches to continue in the foreseeable future. This commitment varies considerably, however, with substantial minorities of staff at all levels being dragged along somewhat unwillingly. Their influence can be seen in the restricted meanings often attached to participation. People's involvement in the new approaches is often 'bought' by the use of developmental funds (for example, the provision of wage-labour opportunities in building check-dams, access roads, cattle-proof trenches or walls). Some members of the FD thereby use participatory approaches in a manipulative way, as a tool to meet their own goals of reducing overt conflict with villagers, involving villagers in protection activities, and reducing biotic pressure on 'core' FD forests.

These Indian FDs thus fall well within the range of situations described by Thompson (1995) and Grindle and Hildebrand (1995). FDs remain hierarchical in structure, with punishments, promotion and other incentives rarely based on transparent criteria. Central elements to this structure (such as the confidential reports, transfers as a mode of punishment and reward, and a bias against specialisation) militate against the possibility of ensuring committed, experienced leadership and trained supportive staff to manage participatory approaches in the medium term. There is no reason to believe that attitudes will inevitably move towards greater participation from the hostile or manipulative positions (Vira 1999). Pressures to be less participatory arise from commercial users of forest products or from pressures to reduce the size or expenditure of the FD and to increase the financial returns. In addition FD staff may need to find unofficial means of supplementing their official salaries. As the incomes of forest users rise, we can also expect that there will be changing popular pressures and changing patterns of forest resource use. Our current view is that FD staff are sufficiently committed to the new approaches that outright hostility and manipulative use of JFM is likely to remain no more than a minority position. But the major requirement of a shift to full support to participatory approaches, and all that this implies, is still some way from being achieved.

Notes

1. The research reported here was supervised by a team from the University of Edinburgh, Scotland and the Indian Council for Forestry Research and Education, Dehra Dun, UP. Research was coordinated by Nandini Sundar: research fellows were Prafulla Gorada and Sagarika Chetty (Andhra Pradesh); Ajith Chandran and Monika Singh (Gujarat); Nabarun Sen Gupta and Shilpa Vasavada (Madhya Pradesh); Abha Mishra and Neeraj Peter (Orissa). Pradeep Khanna interviewed senior Forest Department staff, assisted by Abha Mishra. We gratefully acknowledge a grant from the Economic and Social Research Council (UK) under their Global Environmental Change Initiative: we alone remain responsible for the views expressed here. The title of the chapter is derived from an interview with a senior forest department official in Madhya Pradesh.
2. The interview material is reported as follows: the state is identified by its letter; the broad grade of respondent is identified similarly (S = senior staff, M = middle managers, and F = field staff) and each interviewee in each category has a unique identifying number. Interviews were carried out in May

1997. Structured questionnaires for middle managers and field-level staff were based on those used in AP (Maheshwari and Moosvi, n.d.). The senior staff are almost all from the Indian Forest Service (IFS); they are not the only policy-makers, and one of their complaints is that their proposals can be over-ridden by generalist administrators, members of the Indian Administrative Service (IAS), to whom they report.

3. See also the discussion in Clark (1995). Since 1996, the strong support of the N. Chandrababu Naidu government has changed the nature and political significance of the programme in AP.

4. Some senior officials also claimed that the Arabari experiment was unique not in its methods but in the publicity it has received, and that forest officials in other States have been as innovative but have not boasted about their achievements. The most obvious examples of this are the FD initiatives in Jhabua and Harda in MP (Bahuguna, 1992), and the work of NGOs such as the Aga Khan Rural Support Programme and the Vikram Sarabhai Trust in Gujarat.

5. See, for example, the SWOT (Strengths, Weaknesses, Opportunities and Threats) analyses reported in the reports of workshops organised by the World Bank in Hyderabad in 1994 (D'Silva 1995) and in Agra in 1995.

6. They are less happy with an expanded future role for NGOs, which we discuss further below. They are opposed to the involvement of the private sector in good reserved forests, but not necessarily in degraded forest.

7. Joshi (1998) notes that in West Bengal, field-level staff are similarly supportive of JFM because it reduces the physical attacks they were otherwise facing.

Part IV

Dynamics of Participatory Processes

10
Conflicts Affecting Participatory Forest Management: Their Nature and Implications

Czech Conroy, Abha Mishra, Ajay Rai, Neera M. Singh and Man-Kwun Chan

1. Introduction

Participatory forest management (PFM)[1] is a complex business. Forests provide a wide range of products of subsistence and/or commercial value (direct uses); as well as performing environmental services (indirect uses), and often having religious or cultural significance (non-use values). Even small 'patches' of forest may be used by people from several villages or hamlets; and different sub-groups within a particular hamlet or village may derive different products from the forest. Management of larger areas of forest is even more complex, as they may transcend administrative, political and social boundaries. Given the complexity of forest management, designing PFM programmes or working out how best to support community forest management (CFM),[2] is not easy, and conflicts are probably unavoidable, if not inherent (Anderson *et al.* 1998). The factors giving rise to conflicts need to be better understood, and are discussed in this chapter.

This chapter draws heavily on our knowledge of self-initiated CFM in the state of Orissa, India, where a few thousand[3] communities are managing forests. Their experiences have been documented and analysed in a number of studies.[4] Many communities initiated CFM more than 20 years ago, and hence there is an extensive body of information on what can happen to CFM initiatives over time, and the kinds of conflicts that can arise. Since 1988 the state government has taken a more active interest in CFM, and encouraged CFM groups to join its Joint Forest Management (JFM) programme.[5] In some cases this has generated new forms of conflict between communities and the state.

The chapter is structured as follows. Section 2 describes some of the main types of stakeholders that have interests in forests and their products. It highlights the fact that even within communities there may be different sub-groups with different, and sometimes conflicting, interests. Section 3 describes the various types of conflict that can affect PFM. Conflicts relating to, or affecting, PFM are quite common in Orissa, and occasionally undermine it.

Section 4 considers what implications the various types of conflicts have for external interventions to promote PFM. It describes various measures that can help to identify, avoid, minimise or better manage conflicts. Participatory natural resource management projects have sometimes been weak in recognising and taking account of conflicting interests (Grimble *et al.* 1995). Attention is drawn to the fact that PFM does not take place in a policy vacuum and that a PFM programme may require changes in the macro-environment. The use of stakeholder analysis is discussed, particularly in relation to improving the design of policies, programmes or projects. However, many conflicts cannot be anticipated and taken into account at the design stage, particularly since some conflicts that negatively impact on forest management are primarily concerned with other issues. Thus, it is desirable for PFM projects and programmes to include capacity-building for conflict management as an important component. Alternative Conflict Management (ACM) is an approach that has proved effective in other sectors, and has great potential for managing conflicts in participatory NRM. Section 5 contains some general conclusions.

2. Multiple stakeholders means multiple interests

Table 10.1 lists key stakeholders with an interest (actual or potential) in forests and PFM: it is based on the situation in Orissa, but a similar set of stakeholders would be found in most countries. Stakeholders have been defined as 'any group of people, organised or unorganised, who share a common interest or stake in a particular issue or system' (Grimble and Wellard 1997). They can be at any level or position in society, from global to household or intra-household. A distinction is made between primary and secondary stakeholders. The former are those who depend significantly on a particular area of forest for their livelihoods: they usually live in or near the forest. There is plenty of scope for conflicts, as each stakeholder is likely to have different interests and objectives (in particular, those of poorer and weaker groups could easily be disregarded or marginalised).

Table 10.1 Key stakeholders in PFM: the case of Orissa

Level	Stakeholders
Local on-site – primary	• Management community • Different sub-groups of protecting community (distinguished by class, caste, gender, etc.) • Village leader(s) • Other communities nearby who previously used the protected forest, or who are still allowed limited access to the forest and/or selected products
Local off-site – secondary	• Federation/apex body of protecting communities • Traditional multi-village body • Panchayat
District/forest range	• District Forest Office • Private sector commercial bodies (e.g. NTFP traders, logging companies, organised timber smugglers, mining companies) • NGOs (forest-support, environment, etc.) • Urban consumers of forest products (esp. fuelwood)
State	• Forest Department • Revenue Department • Watershed mission • Ministry of Forests and Environment • Orissa Forest Development Corporation • Tribal Development Cooperative
National government	• Ministry of Environment and Forests
International donor agency	• Swedish International Development Agency

2.1 Communities and conflicts of interest

The term 'Community Forest Management' is widely used. The word 'community' is sometimes taken to imply a group of people living in harmony with each other and with a common set of interests. In India there have been many studies of the impact of JFM, and '[m]ost … have tended to gather aggregated data on overall increases in production of selected species and products from forests brought under JFM … and reached conclusions about the present and future benefits to the "community", "the people" or "the villagers"' (Sarin 1998).

In many villages, however, there are numerous sub-groups: the land-poor and the land-rich; men and women; people of different castes, etc. (Guijt and Shah 1998). The relations between these sub-groups have tended to be neglected in the literature on CPR use, at least until recently (Beck 1994). Yet 'control and conflict over such resources [can be] ... closely tied to power relations [within villages]' (Beck 1994), as is illustrated by the micro–micro conflict example in the Appendix.

Different sub-groups may have different priorities regarding the benefits to be derived from the forest; and hence different objectives and motivations for forest protection. Thus, CFM or JFM may have a differential impact on different sub-groups (Saigal *et al.* 1996; Sarin 1998), and this may be a source of conflict. Some sub-groups may object to the placing of restrictions on the harvesting of products: for example, in Orissa, groups that make a living from selling fuelwood or making bamboo products are sometimes opposed to CFM. Yet few studies of the impact of JFM have examined 'who, within communities and households, has gained and who has lost by class, caste, ethnicity and gender' (Sarin 1998).

The above considerations do not invalidate the concept of a community, but they do mean that the term needs to be carefully defined. The following definition is assumed for the purposes of this chapter:

> a set of people (i) with some shared beliefs, including normative beliefs, and preferences, beyond those constituting their collective action problem, (ii) with a more-or-less stable set of members, (iii) who expect to continue interacting with each other for some time to come, and (iv) whose relations are direct (unmediated by third parties) and multiplex. (Ostrom 1992)

3. The nature of conflicts affecting forest management

A wide range of conflicts has been experienced in Orissa in PFM. Nobody knows exactly how prevalent they are, but they are certainly not unusual. When CFM breaks down or stops, it is usually because of conflict. The majority of CFM initiatives surveyed in our current research project have experienced conflicts that have led to a breakdown (temporary or permanent) of CFM, and/or changes in the protection arrangements. In some cases this has been associated with substantial degradation of the protected forest. It should be noted, however, that the majority of micro–micro level (see below) conflicts are effectively resolved by communities sooner or later.

The relationships between various stakeholders may involve occasional (acute) conflicts, or ongoing (chronic) ones. Some may be readily visible to outsiders, while others may be almost invisible or 'subterranean' (Sarin 1996). Simplifying things somewhat, one can say that conflicts occur at micro or macro levels, and between these levels, and can be classified as follows: micro–micro, micro–macro, or macro–macro.[6] Examples of various types of conflicts in PFM are given below.

3.1 Micro–micro conflicts

Micro–micro-type conflicts can be classified further into four categories (Conroy *et al.* 1999; see Table 10.2), in terms of whether they are within the community protecting the forest or between that community and other stakeholders; and, whether the conflict is directly or indirectly related to forest management. The latter may not always be a clearcut distinction: where there is a history of conflict or mistrust between different stakeholders regarding non-forest matters, there is more likely to be conflict between them in relation to forest management.

The Appendix contains examples of type A and B micro–micro conflicts respectively. The second example has been classified as intra-community, because the two villages involved were *jointly* protecting

Table 10.2 Types of micro–micro conflicts, with examples

	Directly related to Protection	Indirect effect on Protection
Within protection communities[a]	A. One sub-group refuses to abide by protection or harvesting rules	B. Conflict breaks out between 2 sub-groups, who refuse to cooperate any longer in various matters. Forest protection is affected, sometimes leading to a tree-felling free-for-all
Between protection community and other local stakeholder	C. 1 + local stakeholders (e.g. communities, local FD staff, loggers) challenge or do not accept a protection initiative (and may cut down trees in the protected patch)	D. Conflict breaks out between two communities, related to non-protection issues (such as party politics or personal disputes), leading non-protecting community to 'loot' the protected patch

[a] In combined community protection (i.e. involving more than one village or hamlet) each community is classified as a sub-group.

the same patch of forest. Out of 33 CFM initiatives studied in Orissa several had experienced either type A, B or C conflicts. None had experienced type D, although it is possible that in some cases what have been categorised as Type C conflicts may be Type D.

3.2   Micro–macro conflicts

These are conflicts between micro-level stakeholders and higher-level stakeholders, such as government agencies. The relationship between the state and forest management communities is obviously very important in JFM, and can also be important in self-initiated CFM. Various types of conflict may exist or arise between communities and state-level bodies. The legal and policy environment (e.g. regarding tenure, or collection and marketing of non-timber forest products – NTFPs) may have a major influence on the success or otherwise of PFM; hence it needs to be reviewed, and changed if necessary to make it conducive to PFM. Interventions at the micro-level alone may be inadequate. The influence on PFM of the macro-environment for NTFP collection and marketing will now be examined.

3.2.1   *Micro–macro conflicts: the case of non-timber forest products*

Staff of forest departments in India still tend to see timber as the most important product to be derived from the forest, and to regard any other forest products as of minor importance. For local communities involved in forest protection, however, NTFPs are sometimes the most important. This is largely because the benefits from NTFPs are available during the first few years and are fairly certain; whereas the benefits of timber are many years in the future and are likely to be perceived as uncertain. For these reasons, the size of the benefits available from NTFPs may have an important bearing on a community's willingness to become involved in, and to sustain, forest protection (Saigal *et al.* 1996).

Unfortunately, the legal, policy and marketing environments for NTFP collection and processing in India generally operate to the detriment of local people who want to derive an income from NTFPs. In most Indian states 'the marketing environment for realising the full value from NTFPs is constrained by exploitative governmental regulations restricting sale, processing and transport' (Saxena 1997). Often local people have no right to process NTFPs or sell them on the open market. Consequently, although some people, and particularly poorer groups, are critically dependent on NTFP collection for both subsistence and cash needs, the returns are low and it appears that in many cases 'people take it up only for want of any other alternatives' (Saigal *et al.* 1996).

In Orissa, the Forest Act 1972 gives the state the right to exercise a monopoly over any declared forest produce. Almost all the important NTFPs have been nationalised, which means that they can be sold only to government agencies or to agencies nominated by the government. The major institutions set up by the state were the Orissa Forest Development Corporation and the Tribal Development Cooperative (TDCC). Collection rights for a large number of NTFPs have been given to various private sector companies and traders, leading to the creation of private monopolies. Thus, there are major market imperfections (Saxena 1997).

Conflicts relating to NTFP collection and marketing are described in the Appendix. The marketing example illustrates how the policies and practices of state agencies, lack of clarity on the part of state agencies over legal/administrative powers, and a lack of concern about the livelihoods of local communities can seriously damage the financial interests of a community. It highlights the need for sharing arrangements between forest departments and communities to be clearly spelt out and understood at the outset of any agreement.

3.3   Macro–macro conflicts

State governments in India tend to treat JFM as another isolated programme, which they think can be implemented without making any changes in other sectoral programmes (Saxena 1997). However, several aspects of policies, programmes and laws may have a strong influence on the success of JFM.

In Orissa, there appears to be a conflict between the government's JFM programme (and overall forest policy) and the macro-environment for NTFP collection and marketing.

In communities that are not protecting nearby forests there is often a high degree of dependency of some community members on *unsustainable* harvesting of timber and other forest resources, which can serve as a major deterrent to initiation of protection (Conroy *et al.* 1999). These community members perceive that CFM would oblige them to reduce their exploitation of forest resources (e.g. timber, firewood, bamboo), and they were not prepared to accept this because of their high dependency. Community members said that they would be happy to switch to collection of NTFPs as a major source of income if NTFP collection and marketing became sufficiently remunerative for them. Thus, the poor returns available under the government's current NTFP policies and practices are fuelling deforestation, whereas the general policy objective, and a major objective for the JFM programme, is forest conservation.

4. The implications of conflicts for external PFM interventions

Conflicts are liable to occur from time to time, and are not necessarily undesirable. In PFM initiatives that have been promoted by state agencies, such as forest departments, they may sometimes indicate where improvements need to be made (e.g. where group formation had been unsound) and provide an opportunity for change. Interventions by external agencies to support conflict management are not always necessary, as communities may be able to manage conflicts satisfactorily on their own; nor are they always desirable (Warner and Jones 1998).

In PFM, however, if conflicts escalate rapidly they can undermine, almost overnight, several years of community effort in protecting a forest patch. Thus, development agencies should: (1) do what they can to avoid creating or exacerbating conflicts; (2) seek to minimise conflicts when designing PFM initiatives; and, (3) assist in conflict management where local mechanisms are inadequate or non-existent, either by acting as a relatively neutral third party, or by supporting the development of local institutions.

4.1   Designing PFM interventions to accommodate different stakeholders and interests

The formulation of policies, programmes and projects to promote PFM should be sensitive to issues of: differential impacts, including potential winners and losers; political economy; and the likely attitudes and behaviour of different stakeholders.

Stakeholder analysis (SA) can make a useful contribution to the design of policies and interventions in natural resource management. In the NRM sector any policy or intervention is likely to have consequences that bear differentially on different groups and individuals, and on 'society' as a whole: 'unless we know what these differential effects are likely to be, it is impossible to assess the value or worth of that intervention or policy' (Grimble and Wellard 1997). SA attempts to identify winners, losers and 'payoffs'; and to assist the development of 'socially best' policies and interventions (ibid.). Stakeholder analysis can be used: (a) to improve the effectiveness of policies and projects; and/or (b) to address their social and distributional impacts (Grimble and Chan 1995). Its proponents argue that, by identifying potential conflicts between the interests of different stakeholders, it 'helps avoid the unexpected, facilitates good design, improves the likelihood of successful implementation, and assists the assessment of outcomes' (Grimble and Wellard 1997). SA can help to make different objectives mutually compatible by identifying common ground, if it exists, between a number of stakeholders; and hence can assist in the designing of policies and interventions that result in win-win situations (see Table 10.3, below p. 176).

It should be borne in mind that some stakeholders may not want to acknowledge some of their interests. For example, FD staff are unlikely to admit to receiving income from their collusion with timber smugglers: or government officials to receiving money from private traders to whom they have awarded NTFP collection and processing contracts. Such hidden agendas can be brought out into the open by asking different stakeholders to specify what they see *each other's* interests as being, as well as their own, and by asking them to provide supporting evidence based on their experiences.

4.2 Changing the balance of power between the state and communities

It would be extremely naïve to assume that the application of SA will ensure that the interests of weaker groups/stakeholders are respected (Hildyard *et al.* 1998). The stronger stakeholders can be expected to continue to dominate decision-making; and, when conflicts arise, to promote their interests over those of others. Many state agencies may be opposed to giving more power to local communities; and may see it as a threat to established patron–client and rent-seeking relationships (Hobley 1996). There may even be a reverse tendency whereby the state seeks 'to expropriate the initiatives of the people' (Jodha 1990); and, as in India, a history of conflict between certain state institutions and forest-dependent communities over forest resources (Pathak 1994).

Traditional relationships between state agencies and communities are likely to manifest themselves in shared forest management initiatives. This is illustrated by JFM programmes in India, in which FDs (or individuals or groups within them) often have an ambivalent or hostile attitude towards devolving powers to forest protection committees (FPCs) (see also Jeffery *et al.* this volume, chapter 9), and they sometimes unilaterally over-rule FPC decisions, without explanation; and dominate the preparation of micro-plans, which become an instrument by which the FD retains control over the community (Saxena 1997).

Power relations between the state and communities cannot be changed overnight, but in certain institutional and political situations there may be room for manoeuvre within which steps can be taken to promote changes. Three types of measures will now be described.

4.2.1   *Creating a legal or administrative basis for mutual accountability*

In many state JFM programmes in India, FDs have the power to cancel or dissolve FPCs for failing to comply with certain provisions of the JFM resolution or other state rules and regulations. Furthermore, the reasons for the dissolution can be formulated in such a way that the decision does not appear arbitrary (Saxena 1997). The FPCs, on the other hand, are not given any formal rights or mechanisms by which they can bring the FD to account. Thus, the FPCs are accountable to the FD, but not vice versa, making the relationship between them highly unequal. Legal or administrative orders embodying some form of mutual accountability would contribute to a shift in power.

4.2.2   *Creating multi-stakeholder decision-making fora for PFM*

In most countries, PFM involves a major shift from state management of forests to some form of shared management, involving at least two (usually several) major sets of stakeholders. Thus, new multi-stakeholder fora will generally be required (Anderson *et al.* 1998), which should ideally have decision-making powers rather than merely having consultative status. If forest-dependent communities are represented on them, they can: (1) strengthen communities' bargaining power vis-à-vis the state (Vira *et al.* 1998); and (2) help to ensure that negotiations and decision-making will be mutually acceptable to (or at least accepted by) all major stakeholders.

District or division-level fora   For Orissa, it has been proposed that committees be established comprising representatives of CFM groups, OFD and NGOs: one such committee could be constituted for each forest division or each district (Conroy *et al.* 1999). The establishment of this kind of committee or working group is being given consideration by the state government. They would deal with the following kinds of issues (Conroy *et al.* 1999), and could also have a general responsibility for monitoring the performance of PFM initiatives (Vira *et al.* 1998): (1) lack of FD support, either in dealing with offenders or upholding the CFM group's rights; (2) resolution of inter-village boundary disputes over areas of Reserved Forest managed by several communities; (3) alleged involvement of FD staff in timber smuggling from protected patches; (4) undue interference of FD staff in the development or implementation of management plans by communities; (5) concern of FD staff that management plans are not ecologically sound; (6) concern of FD staff over serious deviations from the management plan (e.g. the number of trees being felled by CFM group members);

(7) concern of FD staff that the CFM group is not enforcing protection adequately.

State and national level fora A similar body is also desirable at a higher level to influence the broader enabling environment, including policy and legislation. Ensuring authentic and effective representation of FPCs at this level is more difficult, however.

4.2.3 *Creating and developing forest community apex bodies*

Forest-dependent communities involved in PFM tend to be weak, to function in isolation from each other and to interface with the state individually. Their bargaining power would be strengthened if they could collaborate and take a united stance on certain issues. In Orissa, it is quite common for several communities to work together in CFM, particularly where there is a large tract of forest, and they often form apex bodies to coordinate their activities, assist with conflict management and provide an interface for dealings with the FD (Saxena 1997; Poffenberger *et al.* 1996). Some NGOs in Orissa, such as Vasundhara and the Regional Centre for Development Cooperation, have also encouraged the development of apex bodies, including district-level federations.

4.3 Political economy, equity and conflict at the community level

Most donors and governments promoting PFM initiatives are, at least nominally, committed to benefiting the poorest groups. Since the poor tend to be the most dependent on forest products, they may also stand to lose most from protection – at least, in the short term; and they are the ones whose priorities are most likely to be ignored. Where SA (or any other approach) is used with this objective in mind, the stakeholders selected for the analysis should include *all* those groups, including minorities and the poor, that will be affected in some way by implementation.[7]

In promoting equity the political economy of the situation needs to be taken into account, as power relations within communities cannot be changed easily. Table 10.3 provides a useful classification of strategies in relation to their political feasibility. An attempt to redistribute benefits radically (Type D strategy) may undermine any chance of effective implementation. In practice, therefore, the major opportunities lie with Types B and C strategies. Type B approaches may be the only feasible option where the poor are weak and unorganised, but Type C may become possible where the marginal groups are strong and united.

Table 10.3   Distribution strategies and their political feasibility

Strategy	Rural elite	Rural poor	Political feasibility
A	Gain	Lose	High
B	Gain	Gain	High
C	No change	Gain	Medium
D	Lose	Gain	Low

Source: Adapted from Chambers *et al.* 1989.

The research in Orissa found that conflicts are likely to be minimised where the decision-making processes are transparent and perceived to be fair (Conroy *et al.* 1999). For example, conflict is less likely where all sub-groups are represented on the management committee, where meetings take place regularly, and where records are kept of decisions and financial matters. Misappropriation of funds from CFM by one or more members of the management committee is sometimes a source of conflict between sub-groups, and is likely to be minimised by these processes.

4.4   Developing capacity for conflict management

Some conflicts will occur in PFM no matter how well the programme or policy has been designed. Not all conflicts will be directly related to forest management, but they may nevertheless impinge on it. It is important, therefore, that strong and effective conflict management mechanisms and institutions are available. When conducting SA, any existing ones should be identified and their effectiveness assessed; and, where necessary, new ones should be introduced.

There are various options available for conflict management, the main ones being: force (e.g. adversarial negotiations and legal processes; physical force, public protest), withdrawal, accommodation, compromise (arbitration, trade-offs) and consensus. Force may be necessary for dealing with particular stakeholders and their activities (e.g. timber smugglers, encroachers, poachers), in which case suitable legislation should be in place and the resources needed to enforce it made available to the forest department, police, etc.

4.4.1   Consensus-based approaches

Where different stakeholders are prepared to negotiate peacefully, consensus-based approaches (sometimes called *consensual negotiations or alternative conflict management*) may be best, as they seek to generate mutual gains with the minimum of compromise and trade-off (Warner and Jones 1998). Alternative conflict management (ACM) has evolved primarily from experiences and thinking in peacebuilding and business, and environmental disputes. A summary of the principles of ACM is given in Box 10.1. Further information about the application of ACM in participatory natural resource management can be found in ODI (1998).

Some observers have expressed scepticism about the feasibility and efficacy of consensus-based approaches in natural resource management, arguing that 'consensus on questions of substance ... is highly unlikely or partial and temporary at best' (Anderson *et al.* 1998); while others are more optimistic (ODI, 98). Scepticism seems premature at this point, given the demonstrated effectiveness of ACM or related approaches in other sectors, such as principled negotiation in business (Fisher *et al.* 1997), and the fact that it has hardly been tried yet in NRM. It may be most effective in micro–micro conflicts, where the number of stakeholders is small and power relations between them are not highly skewed.

Two broad types of conflict management assistance can be built into projects and programmes (Warner and Jones 1998), namely direct provision of facilitation or mediation services; and/or training in consensual negotiation, facilitation and mediation skills. The goal of both types of assistance is 'facilitating people to bring about change of their own choosing' (Resolve 1994).

Training can be provided for communities themselves (including apex CBOs), for NGOs and/or for forest department staff (see next section).

Box 10.1 Principles of ACM

- Full stakeholder analysis (including those who might contribute to a resolution and those who might undermine it)
- Cultural differences accommodated in the design of capacity-building and negotiation strategies
- Perceptions acknowledged and then transformed
- Meaningful communication pathways constructed
- A 'level playing field' for genuine collaborative negotiations created
- Rapport built and maintained
- Negotiations focus rapidly onto underlying needs and motivations
- Common ground identified and exploited
- Creative options brainstormed and widened
- Motivations and options re-framed and clarified
- Mutual gains facilitated
- Agreements tested for financial, technical and democratic feasibility

Where there are effective indigenous approaches to conflict management, training in consensual negotiation can build on them.

4.5   Who should manage conflicts?

Ideally, CFM groups themselves should manage micro–micro conflicts affecting them; and it appears that the majority of cases of conflict in Orissa are dealt with effectively by them (Conroy *et al.* 1999). The conflict may be addressed by the forest protection committee itself. Alternatively, there are often traditional institutions or authorities (e.g. village leaders) that play a role in conflict management at the community/village level, and at the multi-village level. However, the Orissa experience shows that their power and influence may wane over time, and often new bodies are required, particularly at the multi-community level.

The evolution and strengthening of *apex organisations* of forest management communities can create new fora for conflict management between member communities. In Orissa there are many examples of such bodies being initiated and developed by the local communities themselves, and they generally identify inter-community conflict management as one of their main functions. NGOs may sometimes have a direct role to play in conflict management. NGOs in Orissa sometimes assist as neutral third parties; and have been able to break situations of deadlock and create an environment for the conflicting parties to come to negotiations.

Forest Department staff may also have a role to play. In Orissa they are frequently called upon by CFM groups managing Reserved Forest to provide third party mediation, usually over conflicts relating to boundary disputes. It should be borne in mind, however, that FD staff are not always neutral parties; and that conflict management may not come easily to them, and may be perceived as an extra burden on their time. For shared forest management programmes there may be a case for creating one or more units specialising in conflict management.

4.6   Conflict-prone PFM implementation issues

4.6.1   *Determining the forest management unit*

JFM programmes in India tend to recognise only legally designated Revenue villages as forest management units, although some Revenue villages are composed of a number of hamlets. The JFM approach is too rigid in this respect, and tends to be conflict-prone. The CFM experience shows that forests are sometimes managed by one or two hamlets, and that other hamlets in a Revenue village may not be involved or may have established a separate CFM initiative (Conroy *et al.* 1999). There may even be forest-related conflicts between different hamlets in the same Revenue village. This suggests that a more flexible approach is needed.

4.6.2 Determining forest management objectives

The objectives of forest management will strongly influence the way in which the forest is managed and the silvicultural practices that are adopted. Different stakeholders may have conflicting management objectives. For example, JFM resolutions refer to the concepts of 'final harvest' and 'major harvest': these terms belong to conventional plantation forestry and reflect the objectives of forest departments. They are alien to most communities, and are not appropriate to the multi-species, multi-purpose forestry practised by CFM groups.

Even within a community, the objectives of one group may conflict with those of another. Sal forests are widespread in parts of Orissa. Sal trees (*Shorea robusta*) can produce several valuable products, notably: good timber, fuelwood, leaves that are widely used to make plates, and seeds from which oil is extracted. If sal was being managed solely for timber production, all the coppice shoots except one would be removed; and the number of green leaves available, particularly at lower levels within the reach of collectors, would be reduced. Thus, women for whom sal plate-making is an important livelihood enterprise would be losers under this management system.

It is important that the management priorities and objectives of all sub-groups are clarified at the outset. In self-initiated forest management in Orissa, as in FPCs established through JFM programmes (Sarin *et al.* 1998), it tends to be the case that an elite group plays the lead role, and formulates management plans without much consideration for weaker ethnic groups (see the first example in the Appendix) or for the interests of women (as in the example above). On the basis of this information it may be possible (though not necessarily always) to develop a management plan that benefits the poorer groups, and ensures that no one group loses out.

5. Conclusions

There are usually a large number of stakeholders to consider in PFM, many with different priorities and objectives regarding forest management. Consequently, there is plenty of scope for conflicts of interest. Generally speaking, participatory forest management initiatives have not given adequate consideration to these issues, although that is starting to change (Anderson *et al.* 1998; Vira *et al.* 1998); and conflicts are common in PFM programmes. Related to this has been the general tendency to ignore power relations between different sub-groups within a given community. There is evidence that the interests of the poorer groups and women have often been overlooked in India's JFM programmes.

Stakeholder analysis can play a valuable role in identifying all of the stakeholders, and in identifying ways of reconciling their priorities and objectives. It is also important that macro-level factors, such as NTFP policies and practices, are taken into account; and, if necessary, revised so as to provide a supportive enabling environment for PFM. Discussions among different stakeholders involved in PFM need to be ongoing and institutionalised, and this is likely to require the creation of new fora.

Some conflicts are inevitable in PFM programmes. It is important, therefore, to ensure that there is adequate capacity to deal with conflicts when they do arise. PFM programmes should, therefore, include provision for capacity development for conflict management, which can take two forms: the creation of new mechanisms and bodies for mediation of conflicts; and training in consensual negotiation, facilitation and mediation skills.

Forests are different from many other renewable natural resources, in that the resource, or a large proportion of it, can be removed virtually overnight. Thus, failure to resolve conflicts quickly may result in local communities losing much of the physical and social assets that they have spent years building up.

Appendix 10.1   Case studies of conflict

1. Adendungri: a type 'A' micro–micro conflict

Adendungri is a village of 139 households in Balangir District, Orissa. It is dominated numerically, as well as economically, by people of the Kulita caste. Kulitas, together with people from some other 'backward castes', account for 82 households (hh); followed by the Mirdha (42 hh), who are tribals, and scheduled castes (15 hh). Protection was initiated in 1968.

In 1973 it was decided that the village should have a temple, and that its construction should have first claim on any income from sale of forest products. (The temple is not completed yet.) The Mirdhas were not directly involved in the initial phase of protection. They had no representation on the temple committee that was also acting as the management committee for the forest: on the other hand, they did not contribute voluntary labour or materials towards the construction of the temple. From the mid-1980s, some of the Mirdha families, who had been living at the fringe of the village, started moving inside it. They then began to feel that they were entitled to equal rights over the forest resources being protected. In the late 1980s and early 1990s incidents of theft from the forest increased, many of them involving Mirdhas.

In 1992 *Panchayat* elections were held. The Mirdhas voted for the candidate from the neighbouring village, and attitudes against them within Adendungri hardened. The newly elected village leader organised group patrolling of the protected forest. Later in 1992 a patrol group was attacked with sticks by a Mirdha group, and two of them were seriously injured. A police case was filed against the Mirdha. Protection broke down, and a free-for-all situation ensued, in which the villagers of Adendungri, and also people from some neighbouring villages, cut and took away almost 50 per cent of the trees. Protection has since been re-initiated, but both groups remain bitter. The Mirdha men say that they have not benefited from the forest being under protection; and that the decision-making body still represents the interests and priorities of the Kulitas.

2. Kesiyapalli and Kulasara: a type 'B' micro–micro conflict

These two villages are located in the Tangi area, south of Bhubaneswar, Orissa's capital. They started forest protection in 1975, when the patch concerned had become highly degraded. Four villages decided to protect, but for practical reasons relating to the size of the protected patch they split the protection responsibilities. Kesiyapalli and Kulasara formed one CFM group, and the other two villages formed another. The combined management system functioned well for nine years, but in 1984 protection broke down.

Causes of the conflict

One factor was that Kesiyapalli wanted to build a road to link up with another road, so that there would be an alternative route to their village for people coming by vehicle from one side. The most direct route to the road they wanted to link up with would have required filling in some of a village pond that they shared with Kulasara. Any other route would have needed to go round the school and hence would have required more labour and money (more purchasing of land). Second, around this time there was a dispute between the two villages over their respective shares of the produce from the pond. Third, there was a *panchayat* election around that time and the two villages had voted for different candidates: this had also increased tensions. Kesiyapalli people had voted for a candidate who belonged to their caste and who ultimately won the election.

The combination of factors caused an escalation of tensions that resulted in the protected patch being severely degraded. Kulasara villagers started cutting trees, people from neighbouring villages soon joined them, and there were soon virtually none left. The Kesiyapalli villagers attempted to save the patch, but things happened so fast that there was not time to resolve the conflict.

3. Micro–macro conflicts in the collection and marketing of NtFPs: some examples from Orissa (main source: Saxena 1997)

Bamboo

Bamboo forests have been assigned to the paper industries, who have been appointed sub-agents of the OFDC. Despite the prescription in the National Forest Policy 1988, that the needs of forestdwellers will constitute the first claim on forest produce, the poor in Orissa have to meet their demand for bamboo by stealing, while the industry receives subsidised bamboo and has first access. Leasing of bamboo coupes to industry may exclude such areas from being brought under participatory management. Alternatively, in one case a community (Paiksahi) that had been protecting a predominantly bamboo forest for several years entered into a JFM arrangement with the FD: but when the bamboo in the protected forest came to be harvested in 1997 it all went to a private contractor, to whom the state government had given a ten-year licence in 1989. Another, more detailed, example of a conflict between a local community and state agencies over bamboo harvesting is given in case 4.

Hill broom

Legislation in Orissa only allows processing of hill broom to be done by the lease holder, TDCC or its traders. Local people can collect hill broom, but they are not allowed to sell the materials freely on the open market, or to bind them into a broom. It has been estimated that a substantial profit could be made, even after paying the prescribed royalty to the Government, if people were allowed to sell their produce on the open market (Anon. 1996, cited in Saigal *et al.* 1996). The state government machinery took action against a women's group that set out to help poor tribal women obtain higher revenue, by binding the material they collected and marketing the brooms: the stocks were seized.

4. A specific micro–macro conflict over bamboo

Ramkhol is a small village with 55 households. Since 1987, it had been involved in protection of its bamboo forest located within the revenue boundary of the village. The management system followed by the villagers was protection, combined with controlled felling of bamboo for *bona fide* needs of community members. In 1995, bamboo in the Ramkhol forest, together with other bamboo in the region, bore flowers. As bamboo dies following its flowering, the villagers approached the local FD officials to allow them to cut down the bamboo in their forest and sell them. Their request for a transit pass, which would have allowed them to transport the bamboo for sale, was refused, as they had not entered into a JFM agreement with the FD. They were assured by the Divisional Forest Officer (DFO) that the matter would be taken care of within fifteen days if they entered into such an agreement.

When the rains started, the villagers became agitated because they knew the rains would destroy the dead bamboo. There was some confusion within the Forest Department about its powers to harvest or sell the produce. In fact it had no power to harvest forest produce on its own, as that rests with Orissa Forest Development Corporation (OFDC); so it needed to coordinate with OFDC. OFDC was in no hurry, and once the rains started, it was not clear whether it could go against the traditional rule of not harvesting and/or transporting bamboo for the next six months.

In January 1996 a federation of about 40 CFM villages was formed in the area, whose committee took up the issue on behalf of Ramkhol. The committee threatened to take legal action against the FD, to claim for the loss incurred by Ramkhol village, if it did not act promptly. After much negotiation, it was decided that the FD should organise the felling and sale, deduct the costs and provide half of the remaining amount to the village. The bamboo was harvested by OFDC, which retained 40 per cent of the net income. The remainder was shared equally by the FD and the community. The villagers felt cheated and believe they should have received about 60 per cent of the net income, not 30 per cent. They regretted not having proceeded with harvest and sale of the bamboo themselves. The FD argued that it receives only a royalty from OFDC, so only that can be shared with the villagers.

Notes

1. PFM is used as an umbrella term covering joint forest management, collaborative forest management, community forestry and, in some cases, social forestry. 'Participatory' has been defined as a process whereby those with legitimate interests in a project both influence decisions which affect them and receive some, or all, of any benefits that mat accrue (ODA 1996).
2. CFM can be described as a system where a community has 'developed institutions, norms, rules, fines and fees to sustain forest resources. CFM systems characteristically involve one or more communities (social group, village) protecting and using a specific forest area' (Poffenberger 1996). While the forest may not be under the legal jurisdiction of the community, 'the community management groups strongly identify with the resource and perceive they have special rights and responsibilities for its management'.
3. Nobody knows exactly how many cases there are. The total number of villages in Orissa is about 50,000: of these, there are probably 4,000–5,000 villages involved in managing natural forests. This is probably more than in any other Indian state or comparable geographical area anywhere else in the world.
4. Valuable general references on CFM in Orissa include: Conroy *et al.* (1999); Jonsson and Rai (1994); Kant *et al.* (1991); Poffenberger *et al.* (1996).
5. JFM can be defined as 'sharing of products, responsibilities, control, and decision making authority over forest lands, between forest departments and local user groups, based on a formal agreement. The primary purpose of JFM is to give users a stake in the forest benefits and a role in planning and management for the sustainable improvement of forest conditions and productivity. A second goal is to support an equitable distribution of forest products' (Hill and Shields 1998). For the background to this programme, see Yadama and Jeffery *et al.* (this volume, chapter 9).
6. Some authors (e.g. Grimble *et al.* 1995) identify two combinations involving micro and macro (i.e. micro–macro and macro–micro), in which the first half of the 'pair' is the active decision-maker, and the second half the passive party. In practice, however, it can be difficult to distinguish the 'active' from the 'passive' party, so we have only used one combination in this chapter.
7. Where SA is being used solely to improve the effectiveness of policies and projects, the stakeholders selected may only include those groups whose interests, resources and position of power imply that they are likely to affect substantially the way in which the project will operate, or fail to operate, in practice.

11

Progress towards Recognising the Rights and Management Potentials of Local Communities in Indonesian State-defined Forest Areas

Chip Fay and Hubert de Foresta

1. Introduction

This chapter reviews current governmental and non-governmental efforts in Indonesia to increase land security and the positive role local people officially play in the management of natural resources inside the 70 per cent of Indonesia defined by the government as a State Forest Zone. It presents a broad-brush description of the dichotomy between local natural resource management systems and the government regulatory framework that favours industrial forestry management. It also describes the characteristics of some of the many indigenous agroforestry systems found in Indonesia and details a successful effort to gain formal government incorporation of one these systems into the state regulatory framework.

The period 1998–99 has been a remarkable time of transition for Indonesia. Indonesians are nurturing a recovery from an economic collapse that brought a 17 per cent contraction of the economy. It is estimated that more than 8 million people lost their jobs. At the same time, the country is emerging from a 32-year period of political darkness. A democratic, multi-party system is slowly taking shape and, as recent elections indicate, the clear majority of Indonesians want a clean break with the corrupt and politically repressive practices of the past.

One area where the reform process is particularly charged is forestry. During the latter half of the Suharto period, legal rights to extract large volumes of high-value tropical timber were awarded to a tight network of Suharto associates and family members. Timber concessions and plantation rights were routinely given out in areas where local communities have lived, managed and depended upon forest resources for generations. Conflicts brought about by the government's refusal to respect local property rights has increased significantly since Suharto's resignation in May 1998, as has the ability of local people to project their problems and demands. As a result, moves towards securing local people's rights are being taken more seriously by the Department of Forestry. Local people's organisations and NGOs, however, say these moves are insufficient in that they focus on granting management rights to areas claimed by local people rather than recognising customary community-based rights or *hak ulayat*.

There are essentially two approaches in Indonesia to increasing the forest management role of local people. The first centres on 'raising the participation of local communities in the management of forest lands'. This is current government policy and programmes to achieve this are being implemented by the Department of Forestry. The second, promoted by local communities, university researchers and NGOs, centres on policy change that shifts the emphasis to government recognition that existing community-based natural resource management systems are the most effective starting point and that land and resource access rights of local people must be secured if community-based forestry is to develop broadly in Indonesia. In many cases this would mean removing agroforestry areas governed by customary community-based rights from the State forest zone and recognising a private, often communal, property right.

The government priority of increasing participation is based upon two important assumptions. The first is that local people are the primary cause of forest degradation or full forest conversion to agriculture. Therefore, any attempt to work with farmers must focus on imposing a new land-use system. The second assumption is that these people have no rights to be on state land at all, whether or not their ancestors had been on the land long before the existence of the Indonesian state. As a result, in some areas of Indonesia, the government's approach to dealing with people living in the forest zone continues to be intimidation and eviction.

Yet a shift is taking place in the government's approach. The Department of Forestry policy no longer treats local people in the forest zone as liabilities alone. They are now seen as liabilities that must and can become assets in government efforts to increase timber production and rehabilitate degraded forest lands.[1] The result is increasing openness to involving local people in predefined Department of Forestry and forest industry activities. A more detailed description of the Indonesian setting for community-based natural resource management and specific Department of Forestry community-oriented programmes are found in sections 2 and 3.

The non-governmental approach has been developed by a coalition of lands rights advocates, university and international social science and policy researchers, and community organisers.[2] It is based upon evidence that in many areas of the state-defined forest zone, local people have ecologically and economically sound natural resource management systems. These groups claim that if government hopes to promote sustainable community-based forest lands management, it must begin with and learn from what people are already doing in these areas. Legal analysis carried out by these groups also challenges the Department of Forestry's position that people living within the State Forest Zone do not have rights to their lands and land-use systems.

Bridging these two approaches is a growing group of government and non-government individuals and institutions that are working to develop solutions that meet the needs of local people for natural resource security and for the Department of Forestry to meet its objectives of forest resource development and the protection of forest functions.

Section 4 provides important detail to what the non-governmental approach is based upon. It describes the variety of indigenous agroforestry systems that can be found in Indonesia and then focuses on the Krui agroforest system found in Lampung province, Sumatra. Non-governmental efforts to gain official recognition of these agroforests recently succeeded in convincing the Minister of Forestry to move significantly beyond the Department's conservative approach to working with local people. In January 1998, the Minister created a special management classification that recognises the ecological and economic benefits of these indigenous agroforests and secures the rights of the Krui agroforestry farmers to protect and develop these systems. Section 5 of this chapter discusses the significance of the new classification and potential for its extrapolation to other areas of Indonesia. The final section provides some reflection and conclusions.

2. Community-based natural resource management: the Indonesian setting

Numerous agroforestry studies have revealed that economically productive and ecologically sound community-based agroforestry systems

are found in many areas of Indonesia that the government has classified as 'state forest lands'. According to the Indonesian government,
these lands are under the sole jurisdiction of the Department of
Forestry and access to natural resources found and developed there is
heavily regulated. Over the past 20 years, Indonesian national development priorities have emphasised large-scale natural resource extraction
from these areas, particularly timber from natural forests. The framework for distribution of concession rights to natural forests has been
highly political. The hundreds of corporations who have received these
rights have enjoyed windfall profits from the rapid mining of timber
while government efforts to regulate extraction and promote sustainable levels of harvesting have largely failed.

While Indonesia had about 152 million hectares of healthy forest in
1950, today less than 95 million hectares remain, making Indonesian
deforestation rates among the highest in the world. Extraction levels
estimated by the World Bank may exceed 40 million cubic metres per
annum, leading to a level of deforestation that approaches 1 million
hectares a year.[3] According to an internal 1995 Bank report, if current
levels of extraction continue, Indonesia will become a net importer of
timber by the year 2013.[4]

But major changes in forestry regulation in Indonesia are currently
being discussed within the context of sweeping economic reforms proposed by the International Monetary Fund. Formal and informal forms
of taxation and politicised regulatory and marketing systems are under
attack.

One area where reform is particularly necessary is in the system of
allocation of access to forest resources. One of Indonesia's greatest
sources of social conflict originates from this system. Under current
regulations, millions of rural people who live inside the 70 per cent of
Indonesia classified as State Forest Zone are considered illegal occu-
pants on land they have farmed and lived upon for generations.
Industrial concession maps continue to be drawn, logging licences and
tree plantations continue to be awarded, and the presence of local villages and their systems of natural resource management, located in the
forest zone, continue to be conveniently ignored. On several occasions
over the past three years, farming families planting within watershed
forests have been forcibly evicted by the Department of Forestry.[5]

But long-term communities living within the state-defined forest
zone are not without rights recognised in Indonesian jurisprudence.
These communities can argue that their forests and their lands are private, and that administrative procedures that classified them as state

forest, violated their rights. The Agrarian Law of 1960 recognises the right of local communities to continue to manage their forested lands under local resource management systems and customary law. According to this law, local communities are entitled to have their rights legally recognised, registered and honoured by government. The spatial planning law of 1992 and law number 10 of 1992 on vulnerable communities both recognise community rights to participate in defining their territories; law number 10 goes as far as recognising community rights to cultural autonomy, including control over their natural resource management systems.

These laws also establish the rights of local communities to know what development activities are being planned and implemented in their areas and to participate fully in the social and environmental impact review of these activities. Yet, in a country where the integrity of the legal system is questionable, having these rights and actually securing them are quite different. The Department of Forestry cites the provision in the Basic Forestry Law that states, traditional (Adat) rights, particularly to land, will be recognised only when these rights do not conflict with 'national interests'. This law also clearly nullifies the existence of 'community forests' (*hutan marga*), the assumption being that these forests are natural and therefore under the patrimony of the state. There are also numerous provincial level decrees that effectively nullify these rights.

The responses of local communities in the forest zone vary. Some villagers who feel they have exhausted their efforts at peaceful dialogue with local government have taken direct and aggressive action against companies that have entered their lands. On several occasions over the past three years, Dayak communities in Kalimantan have burned base camps of timber plantation companies that have ignored local land rights. In other areas, local communities continue to negotiate with local and national government to gain recognition of their rights to their land and agroforestry systems.

Over the past several years, a network of Indonesian academics, human rights and environment activists, international researchers and government officials who are committed to resolving social conflict in the forest zone has emerged and has had measurable impact. Momentum towards increasing the role local people play in forest management and the development of community-oriented programmes has increased and the balance towards greater commitment to local people is shifting. This change can be traced directly to Forestry Minister Djamaludin's commitment to promoting this change.

The Department of Forestry currently has two programmes that focus on meeting the needs of local people inside State forests: Pembinaan Masyarakat Desa Hutan Terpadu (PMDH); Hutan Kemasyarakatan (HKM). In addition, in January 1998 the Minister of Forestry signed a decree that secures the rights of several thousand agroforestry farmers inside the forest zone in Krui, Lampung. This 'Zone with Distinct Purpose' is intended, by the Minister to be a model for securing local rights for indigenous agroforestry farmers who are managing their forests within state forests.

While each of these approaches has their weaknesses and Indonesia still remains behind its ASEAN neighbours in facilitating the flow of benefits to forest communities, it is fair to say that commitment within the Indonesian Department of Forestry towards working with local people has never been higher. But this commitment tends to be found at the higher levels of government, while appreciation for the need for greater community involvement at provincial and district levels remains limited.

In response, universities and NGOs in Indonesia are in the process of developing a two-stage approach to promoting secure land tenure for communities that hold traditional rights. The first entails work within the state regulatory framework and supports the granting of limited use and management rights to local individuals or communities. This responds to the immediate need for halting the conversion of village lands to large-scale forest concessions and plantations while at the same time supports sound management of these areas according to community-level practices. It provides tenure rights to the agroforestry system but the land, according to the government, continues to be controlled by the state. The second stage is a long-term legal and political struggle by local people to gain state recognition that their lands have been misclassified as state forest zone. Many non-governmental organisations have adopted the 'short-term strategy' to securing local resource rights in preparation for the 'long-term strategy' to gain full recognition of traditional land and natural resource rights of forest-based communities.

3. Government efforts to increase the role local people play in forest lands management

When President Jusuf Habibie appointed his Reform Cabinet in June 1998, he chose Dr Muslimin Nasution to be his Minister of Forestry. During the later years of the Suharto government, Minister Nasution had been a senior official in Indonesia's powerful planning agency (Bappanas). Prior to that had been a high level administrator in the Ministry of Cooperatives.

The call 'Forests for the People' featured prominently in the new Minister's early speeches as he laid the groundwork for developing a populist image. The centre of his reformist approach has been a strategy of redistribution of benefits derived from forest resources. The Minister challenged the close partnership between government and the forest industry that during the Suharto period resulted in widespread corruption and mismanagement of forest resources. Allowing cooperatives to manage forestlands, reductions in the area that forest concessions can manage, and a requirement that a portion of all forestry companies that have government-awarded concessions be owned by a cooperative are the initial actions taken by the Minister to promote his redistribution strategy.

Other actions taken were in line with requirements set forth in Indonesia's January 1998 agreement with the International Monetary Fund. These included, among others, the auctioning of forest concessions, the fixing of performance bonds, extending the period of timber concessions from 20 years to 70, the implementation of a resource rent tax, and the reduction of taxes on sawn timber and rattan to 10 per cent ad valorem.

Reform efforts have also been extended to the government's nascent community forestry programmes. The following is a brief description of the two programmes the Department of Forestry is developing, the first being focused on Javan and the second throughout the Outer Islands. Included are examples of interventions by non-governmental actors who are promoting the creation of new experience on the ground and improved policies that strengthen the political and economic position of local people.

3.1 Pembinaan Masyarakat Desa Hutan Terpadu: Java (PMDHT) (integrated forest village development)

This programme also known as the Java Social Forestry Programme, is the latest effort of the State Forestry Corporation, the parastatal forest corporation responsible for the management of state forest lands on Java, to increase the participation of local people in tree plantation development. It is designed to share responsibility of forest community development with local government. The social forestry management system is based on alley cropping. Farmer participants assist in the development of the timber plantation and are allowed to grow crops in between the trees until the time the tree canopy closes, generally a period of between two to five years. This system, known in Indonesia as *Tumpang Sari*, originated during the Dutch occupation and has developed slowly to provide more benefits to local farmers, such as increased spacing of the main tree species.

The programme is currently experimenting with new benefit sharing schemes by increasing the amount of land available to farmer participants for non-timber products. In each of Java's three provinces, pilot areas have been developed where 20 per cent of a timber plantation is handed over to farmers to plant trees of their choice. All benefits from these trees, with the exception of timber, go to the farmers.

3.2   The Hutan Kemasyarakatan Programme (HKM):
(Community Forestry Programme, Outer Islands)

In terms of actual implementation, this is the government's most advanced effort to increase the participation of local communities in the management of state forest lands. The programme's 1995 Ministerial Decree created the first formal Department of Forestry effort at developing a social forestry programme. The programme's initial framework was far more restrictive than the community forestry programmes of Indonesia's neighbours in South and Southeast Asia. During early stages of implementation, HKM participants had no rights to the timber from trees they planted. Like the State Forestry Corporation social forestry programme on Java, only non-timber products (mostly fruit), benefit local people.

Since 1995, numerous Indonesian and international non-governmental organisations have collaborated in an effort to broaden the scope of the HKM programme. The position of these groups has been to increase participation of local people in decision-making, particularly on the composition of the agroforestry systems, and to receive assurances that trees could be harvested by those farmers who planted them.

NGO recommendations on improving the HKM programme centred on the need for the Department to address what was viewed as an imbalance between the programme's rehabilitation objectives and the objectives to increase the welfare of programme participants and reduce conflict between the government and forest dwellers. International Centre for Research on Agro Forestry (ICRAF) staff questioned why wood production was not included in the programme objectives given Indonesia's increasing need for timber. Critics also highlighted the need for the programme to respond to the need of participants to have agroforestry products that could close the gap between the time the multipurpose trees are planted and when farmers can harvest. Horticultural crops were recommended, as well as cacao and coffee that produce output more quickly than most fruit species.

On 7 October 1998 the Minister of Forestry signed ministerial decree 677 (SK 677). It sets out the framework for the second generation of the Hutan Kemasyarakatan (HKM), or community forestry programme. This programme is the government's most advanced effort to increase participation of communities living inside the state-defined forest zone in the management of forest resources. On balance, the new framework represents a significant improvement over the previous programme.

There are four main areas of improvement:

1. The process of developing the policy was, at least through most stages, open, transparent and participative. Several non-governmental organisations and university staff were deeply involved in the conceptualisation of the new framework as well as in the actual drafting of SK 677 (the weaknesses/limitations of this process discussed below).
2. The programme allows the harvesting of both natural and planted timber, unlike the original framework that allowed only for the harvesting of non-timber products.
3. The time limit of the contract awarded to participants was increased from 20 to 35 years and made renewable.
4. The programme is defined by a set of internationally recognised community forestry principles. The two most important are that local communities are the primary actors and that the forest management system for project areas can be based upon existing community-based forest management practices.

Regrettably, the further the process evolved the further the Department moved away from many of the basic principles. Clear contradictions appeared in the final decree. The most blatant was that all community organisations must take the form of cooperatives, ignoring a central principle stated earlier in the decree that the community must define their own organisations. Other contradictions have emerged in the drafting of the implementing and technical guidelines. The tendency has been that Department staffs, when tasked to draft such guidelines, revert back to familiar, prescriptive approaches that run counter to the objectives stated in the community forestry policy framework.

Overall, the greatest weakness of the new framework is its scope. Given the prescriptive and still heavily regulated approach, it is likely that the programme will only be implementable in a few, very limited areas (appropriate for a small cross-section of circumstances in the state forest zone). These would be areas where communities are, as the programme requires, prepared and capable of forming a cooperative and fulfilling the programme reporting requirements. There is also the important consideration of land rights. Many if not most communities inside the state forest zone believe, and can often demonstrate that they have rights over areas that precede state delineation of their lands as national forest. It is likely that these communities will not be satisfied with being awarded conditional rights over areas they claim as their Adat lands.

Yet, in order to place these reforms in perspective, the development of SK 677 may well have been the most participatory forestry policy development process ever undertaken in Indonesia. It was, right up to the time it went to the Minister, an open and transparent process. No group or individual that requested to participate and assist was refused. It was also, perhaps, the best working partnership ever, between Department staff and outside policy groups. A shared vision did emerge, particularly with mid-level forestry staff responsible for drafting during the initial stages.

4. Non-governmental efforts to secure local rights over indigenous agroforestry systems: the Krui experience

Indonesia has a full range of non-governmental organisations active in community development, human rights and forest policy advocacy. Some of these groups are attempting to assist the Department of Forestry implement and improve its community-oriented forestry programmes while others choose to avoid collaboration with government and concentrate on working with local communities to demand that their land rights be recognised. Both these groups have recognised the limitations of the two government community forestry programmes, particularly in that they do little to address community rights that existed prior to the creation of the state forest zone or to recognise the economic and environmental benefits of many community-based agroforestry systems.

Over the past three years, there has been a convergence of these two groups of NGOs. Some groups who work closely with government are becoming sceptical about government commitment while some of those who have not worked with government are coming to terms with the reality that in today's Indonesia, if local communities are to gain some protection against the conversion of their lands to large-scale agriculture or plantation forestry, it will have to be the government that recognises and protects their rights. Orbiting around and at times collaborating with these groups are university and international social science and forestry researchers who generally tend to be sympathetic to the plight of local people living in the state forest zone. When these non-governmental institutions and individuals collaborate and focus on a specific issue and specific place, their impact can be measurable.[6]

One initiative that began in 1994 is *Sistem Hutan Kerakyatan* (SHK) a network of researchers and communities promoting the recognition of existing indigenous agroforestry systems. The SHK (forest management systems of the people) network now has an impressive amount of information on existing systems and policies that serve as supporting or constraining the development of these systems.

Members of this network also work closely with a growing number of NGOs who are training local people to map their ancestral lands and natural resource management systems. In several parts of Indonesia these community-generated maps have been used by local people to articulate the boundaries of their lands and territories as well as the sophistication of their land use systems. Some communities have been able to incorporate their maps into the official spatial planning process of local government using the national law on spatial planning as their entry point. This law stipulates that local people must participate in land use planning at the district and village level.

In 1994, local non-governmental organisations, universities and other international researchers that have been studying the ecology, economics and socio-cultural aspects of the Krui agroforests decided to form a research consortium. Some of these groups were also active in the SHK and community mapping networks. The objective was to foster collaboration and respond better to the needs of Krui agroforestry farmers.[7] An important motivating factor for these groups was the concern of local farmers that approximately 29,000 hectares of the Krui agroforests are located within the State Forestry Zone. The most serious implication of this was that a forestry company held the government-awarded concession that covered these lands. This company therefore, held the right to manage this area, including the possible harvesting of an estimated 3 million commercially valuable trees planted by Krui farmers.[8]

Compounding this threat were a number of oil palm companies that, with local government support, began in 1996 to encroach upon the Krui agroforests areas. In mid-1996, one company clear-cut dozens of hectares of community-planted damar agroforests just south of the Krui area.

Responding to requests for assistance from local villagers, the research consortium began working with Krui farmers to literally place their agroforestry systems on the map and to articulate the environmental and economic benefits of their land-use systems. Research and community organising efforts have produced numerous maps and detailed description and analysis of the Krui agroforests (see Appendix). In June 1997, the consortium initiated a dialogue with government concerning the status of their lands, organised field visits from key government officials and held a two-day conference where research results were presented and the status of the land was discussed. The results of these activities were reported to the Minister of Forestry, together with a request from many villages to have the border of the forest zone changed in order to remove the state forest zone from within Krui agroforests.

This request and the results of the workshop set into motion a process that led, several months later, to the signing of a Ministerial Decree that creates the special forest-use classification (KdTI). This decree, for the first time in Indonesia, recognises the rights of community-based agroforesters to control, maintain and develop their forest management systems within the State Forest Zone. The KdTI was informed by the Ancestral Domain classification experience in the Philippines. At the Minister's request, ICRAF policy research staff working on land tenure issues in the Philippines, translated Philippine community forestry and Ancestral Domain regulations into Bahasa Indonesia for review by the Indonesian Department of Forestry legal staff.

The classification is unprecedented in that:

1. It sanctions a community-based natural resource management system as the official management regime within the State Forest Zone.
2. The Department of Forestry allowed non-governmental organisations working with local people to be directly involved in the drafting of a forestry decree.
3. It allows the harvesting of timber from within the State Forest Zone by local people.
4. It allows the limited harvesting of timber from within a watershed, provided the watershed functions are still met.
5. It devolves the management responsibility of State Forest Lands to a traditional community governing structure (*Masyarakat Hukum Adat*).
6. It is a right provided without a time limit.

The KdTI is based upon a right that will be provided by the Department of Forestry. It is a right that can be accepted or refused by local farmers. Currently, members of the research consortium and local government are organising a process of consultation with village leaders to explain the opportunities and constraints the new classification provides the Krui farmers.

The Minister's action on the Krui agroforests is a breakthrough in efforts to promote, secure and develop community-based natural resource management systems in Indonesia. While designed strictly for the Krui agroforests and based upon a premise that the land is state forest land, the implications for the thousands of villages within the Indonesian State Forest Zone who also seek a tool to gain resource security could be profound. At the Minister's request, copies of this decree were sent to human rights and environmental groups concerned about resource insecurity of people living within the State Forest Zone. It is his hope that this approach will be replicated in other areas of Indonesia.

The process of drafting the ministerial decree was long and complicated. ICRAF was specifically requested by the Minister to assist his staff to develop the concept. In turn, ICRAF, with the blessing of the entire research consortium, brought into the process staff from Latin, an Indonesian Forestry NGO and member of the consortium. A working group was formed within the Department with ICRAF and Latin as members. The actual drafting took place over a period of six weeks. During this time proponents of both perspectives discussed earlier in this paper met head to head. Many mid-level Department staff held strongly to the position that any activity in the Krui area must fall within an existing Department programme, most likely HKM. Some Department staff also insisted that the area in question was no more than 7000 ha, although satellite imagery shows approximately 29,000 of agroforests within the forest zone in Krui.

Over this period eight versions of the decree were drafted and four of these were reviewed by the Minister himself. In the end, with few exceptions, the technical input provided by the research consortium through ICRAF and Latin and supported by key Department staff was accepted by the Minister. The management of the KdTI area can be based upon what farmers are already doing, the duration of the right is open-ended making it inheritable, and limited harvesting of timber from both the production and watershed forests is permitted. The results far exceeded what consortium members expected when initial discussions began with Department staff concerning the shape of the Ministerial Decree.

The creation of the Krui special use zone in 1998 was a victory for those in Indonesia who were encouraging the Department of Forestry to use what is already being done by local farmers as the basis or starting pointing for community-oriented forest management. Following the signing of the decree, the challenge centred on assuring that Krui farmers had the information they needed to decide whether this right meets their needs for resource security and doesn't encumber their efforts to gain full recognition of their land rights.

During initial discussions in two villages, many farmers were determined to stand by their position that the border of the State Forest Zone be moved so their agroforests no longer fall within the jurisdiction of the Department of Forestry. One way of achieving this is for farmers to work through the provincial land-use planning process that determines, in co-ordination with the Department of Forestry, which lands are classified as agriculture and which as forestry. Land classification is reviewed every five years. There is also an emerging process from the Bureau of Lands Registration that local government can recognise and register customary rights of local communities with the intention of securing rights. This process was set in motion by a 1999 decree from the Department of Agraria.[9]

The initial response from many Krui farmers was that since the right provided by the Department does secure their agroforestry systems from outside intervention and negates the right of the forestry company to harvest locally owned trees, the new classification is clearly an improvement. But political changes that followed the Suharto resignation have led Krui leaders to reassess their approach to gaining land rights.[10] The option to gain security through the new classification still exists but remains unacted upon by the Krui communities. Meanwhile Krui leaders continue to explore their preferred option of redrawing the state forest boundary. The organisations involved in assisting in the development of decree continue to assist Krui communities to gain full recognition of the rights over their agroforests.

5. Extrapolation of the Krui experience

While the Krui farmers continue to study and strengthen their options for gaining rights over their lands, forest policy researchers, land rights activists, and sympathetic government officials continue their efforts at developing a national policy on the recognition of Adat rights inside of state forest areas.

The Krui classification served to significantly broaden the parameters of discussions within the Department of what is possible. As a result, Department of Forestry, with assistance from the then retired Minister who signed the Krui decree and ICRAF policy research staff, initiated the drafting of a regulation aimed at recognising Adat areas inside the forest zone. The intention can be described as taking what was determined to be possible in Krui and applying it throughout the forest zone.

This group met with the Minister of Forestry and outlined a two-track approach to dealing with community forestry, the first centring on the existing community forestry programme and the second and complementary approach, centres on a Krui-type arrangement for Adat communities with proven forest management capacities. The Minister accepted these recommendations and requested that his senior staff works with this team and other interested NGOs to develop what, six months later became known as the 'draft Adat decree'.

The process of developing the draft Adat decree is ongoing. A small team that included government and non-government members wrote a first draft. The head of the association of timber concessionaires (APHI) was an active member of this team. His said his participation was prompted by the need for logging companies to know exactly who are the communities, within and around their concessions, that have Adat rights. Following the change of government on May 1998, and Ministerial pronouncements of 'Forests for the People', numerous logging companies have been besieged by communities demanding the removal of concessions from their lands and compensation for resources taken and destroyed. It is not uncommon for one group to demand compensation for a given area one day, and another to demand compensation for the same area the following day. This has led to a situation in many concessions and tree plantations that industry people are referring to as anarchy.

The first draft of the Adat policy, or 'draft zero' as it is referred to in Indonesian to emphasise that there has yet to be any public comments on its contents, was distributed in May 1999. Its completion marked the beginning of a complicated process of having the draft concept work its way through the forestry bureaucracy while at the same time being open for public scrutiny. The Minster himself, having been criticised for a lack of transparency in policy development, insisted that discussion on this policy be open for broad public participation. Meanwhile, the drafting team was asked to prepare an accompanying 'academic draft' that details the technical and legal justifications for such a new policy. This draft was completed in June 1999.

'Draft zero' attempts to deal with the most difficult questions that arise when a government makes a good faith effort to recognise the property rights of communities who have claims that predate the existence of the state. The first is just exactly what is an Adat community? The second is what are the procedures for the government to recognise an Adat community? Third, what are the rights that such communities have that can be recognised by government and how can these rights be delineated? And fourth, how does the government deal with conflicts that arise from overlapping rights, particularly in areas where the government has already awarded rights, such as logging concessions and timber plantations?

The drafting team studied carefully how other countries have dealt with government recognition of indigenous rights, particularly to land. Of the countries looked at, the Philippines offered experience that most closely resembled Indonesian conditions. In 1993, the Philippine Department of Environment and Natural Resources, the agency that manages areas classified as public forest lands, developed the Certificate of Ancestral Domain Claim (CADC). The CADC is a special certificate that is issued to Adat (ancestral) communities who have reasonably demonstrated their claims over classified forestlands. While the classification does not go as far as legally recognising community-based property rights, it does provide Adat communities exclusive and open-ended rights over areas they claim as ancestral. Between 1994 and 1998, 2.5 million hectares or close to 20 per cent of the Philippine forest zone were classified as ancestral areas.[11]

The Philippine CADC experience offers Indonesia an interesting point of reference. Similar to the CADC, the initial concept of the Indonesian Adat policy outlines a process by which the difficult questions and problems would be answered and addressed. The draft definition of an Adat community is taken from the government regulation on Krui, mainly because is highly inclusive and had already been accepted by the Department of Forestry. It simply states that an Adat community is 'a traditional community still bound together in association, having Adat institutions, customary law that is still adhered to, a territory defined by customary law, and whose existence is affirmed by the community itself together with government'.

The draft also calls for the creation of a permanent commission at the national level and a commission at the district level (*Kabupaten*, layer of government one step below the province). Both would be made up of government and non-government individuals. The national commission, based at the Department of Forestry would develop criteria for how a community would gain government recognition as an Adat community. The commission at the district level would be formed by the *Bupati* or district head and provide a recommendation as to whether a community that requested recognition met the criteria determined by the national commission.

When a community defines itself as an Adat community in the state forest zone and gains recognition from the government, the next question is how to shape and secure their property rights in the forestry context. At this time the Department of Forestry still has legal jurisdiction to determine whether such a community is managing the natural resources within their area in a way that meets Department approval. Therefore, any right given to an officially recognised Adat community is conditional. Expecting a community to demonstrate they are managing their resources sustainably may be unreasonable. Many of those developing the Adat policy hope that, at minimum, local Adat communities must merely demonstrate they are not harming their environment. The burden of proof that they are should be in the hands of government. Criteria for sustainable management are not clear, as they are not clear in most forest management situations throughout the world, making this condition one of the more difficult in the Adat recognition process.

6. Conclusions

The process by which forest policy in Indonesia is changing is closely linked to the larger process of national political transformation. In such a dynamic time, determining strategies and priorities for specific reforms, for example addressing the complicated question of property rights in areas delineated by the government as State forests, is a challenging task. Discussion among those involved in the process, government and non-government, often centres on defining realistic expectations on how much change can take place in a given period of time. This comes back to the earlier discussion in this paper concerning short and long term strategies for change. When looking at state forests/Adat questions concerning property rights, the short-term priority, in the minds of those contributing to the development of the Adat policy, is for Adat communities to be empowered to articulate where their territories are within the state forest zone and to gain security of tenure over these lands and agroforestry and forest management systems, thereby preventing other rights from being awarded over their areas. In addition, official recognition of Adat communities should strengthen their ability to protect their lands in situations where other rights, such as logging concessions and timber plantations, have already been awarded and conflict exists.

Yet, the weaknesses of the draft Adat policy must also be recognised. The most important is that all land recognised as having legitimate

Adat claims, would still fall within the boundaries of the state forest zone. It therefore accepts the flawed process by which the state forest-lands were demarcated and determined. Following the open dissemination of the draft policy for comment, much debate has taken place over this question. Critics agree that developing procedures by which Adat communities inside the forest zone can be formally recognised is important, but the process should explicitly include steps by which Adat lands that should never have been classified as state forests can be declassified and *Hak Ulayat* or communal rights recognised. Such processes arguably already exist. The Department of Forestry has long had a procedure of creating enclaves inside the forest zone. These are areas where, according to the Department, other rights exist and there is no clear ecological justification that the land in question serves a forest function. But this procedure is rarely employed. Department legal experts say only areas with full land titles can be excised in this way from the state forest zone and it is uncommon in Indonesia for agricultural or any villages lands to have land titles. Ownership and use rights are most often recognised and registered by local government and transfer takes place through official deeds of sale.

Another emerging opportunity for Adat communities to have their rights over their lands and forests recognised comes through the process of redefining the state forest boundaries. The Department of Forestry has invited public involvement in the development of participatory forest boundary setting procedures, a process intended to reassess where the state forest land should be and where it should not be. Adat lands that serve no forest environmental function and are no logger forested, can be excised from the state forest through having the forest boundary drawn in a way that places these territories outside the forest zone (as opposed to being an island of land inside, as an enclave). This process, has the potential of being less difficult than creating islands of Adat territories as enclaves within the state forest since the tendency of forestry planners is to completely separate community lands and activities from the state forests.

Finally, a process of identifying and officially registering Adat lands is emerging from the Bureau of Lands with the Department of Agraria. Ministerial decree No. 5/1999, or Guidelines to Resolve Adat Communal Rights Conflicts was signed by the Minister of Agraria in March 1999. This decree sets into motion a process that, similar to the Ministry of Forest policy initiative on Adat, will determine criteria for the recognition of *Hak Ulayat*. The main difference is that the BPN will accept the registration of Adat lands and treat them as a communal and non-transferable right, unlike the forestry classification that would provide only a management right.

While not completely satisfied with this new opportunity, some Adat leaders and NGOs have decided to test this process and determine what form of recognition can be gained. The foremost question is what happens in the overlapping areas? The state has already given out 65 million hectares to the timber industry; 15 million to plantations and 48 million hectares are set aside as protected forests including national parks. Added to this list are 482 mining concessions and transmigration areas.

While these questions are being debated, a new forestry law is before the national parliament. The process of drafting of this law has been long, complicated and as a result, contentious. The current draft favours a distribution of the benefits of forest exploitation with a greater share going to local government. Debate continues over how the law will deal with the status of Adat territories within the State forests. Yet, it is likely that opportunities for separating Adat lands from the State forest zone will continue to develop and that natural forests and areas of environmental importance, such as critical watersheds, will remain within the jurisdiction of the Department of Forestry. Adat rights over such areas may well be awarded through a process such as the special classification offered by government to the Krui agroforestry farmers.

Appendix 11.1   Indonesian indigenous agroforestry systems

Until recently, only two 'indigenous' agroforestry systems were officially recognised and referred to as national forms of sophisticated agroforestry developed by people and forestry services: *'tumpangsari'*, an Indonesian version of the taungya system, and *'pekarangan'*, the Javanese homegarden, which is said to be one of the most sophisticated homegarden systems in the world. Apart from these, some systems are occasionally mentioned: the *'talun/kebun'* system of West Java which is formed of an alternation, on the same piece of land, of perennial crops (bamboos, fast-growing tree crops, fruit trees, which form the *'talun'* phase) and of a mixture of annual crops and seedlings of perennials (the *'kebun'* phase, which is a rejuvenating phase of the 'talun'). In Java, almost every piece of land is planted with a mix of perennials and annuals: most dry fields include trees either as true components (coconut with maize) or as borders (teaks, mahogany, rosewood, in East and Central Java). Trees are also commonly associated with irrigated ricefields either on dikes or along roads, and these are useful trees which hold a real role in the agricultural system.

Most of these systems are clearly agroforestry combinations, meaning associations of trees and agricultural crops, but due to the relatively small number of components included, these types of agroforestry associations may be called 'simple agroforestry systems'. In these systems, the mature stage can never be assimilated to a diverse forest. For common observers, forest vegetation bordering agricultural areas is often misunderstood as a mix of 'virgin' and 'degraded' forest. But for experienced agronomists or foresters with a minimum knowledge of botany, or for anyone who can just ask the farmers in these areas, it will soon appear that these are not patches of 'natural', unmanaged vegetation, but in most cases diverse tree gardens. A whole facet of Indonesian agriculture has evolved around traditional forest resources such as fruits, spices, as well as forest material and other commercial resources. Integration of these resources in agricultural lands have gradually shaped original agroforestry systems in which common domesticated tree species of tropical gardening, such as fruit trees, rubber, cinnamon, and coffee, are associated with forest trees (Michon 1985).

The kind of agroforestry combination which is not immediately apparent because it takes the form of a 'forest' both in physiognomy and in function, can be called 'complex agroforestry system' or more simply 'agroforest' (Mary and Michon 1987; de Foresta and Michon 1997). These are successional agroforestry systems in which a high number of components (trees as well as treelets, lianas, herbs) are associated, and physiognomy and functioning of these are often close to those observed for natural ecosystems, either primary or secondary forests. Even though agroforests do not necessarily exhibit an association between agricultural crops and forest trees, they represent what can be considered the heart of agroforestry, where forests and agriculture meet, where forest structures and agricultural logics intersect.

Indonesian agroforests are far from being anecdotal in terms of regional and national economy: they provide about 70 per cent of the total amount of rubber produced in the country, at least 80 per cent of the damar resin, roughly 80–90 per cent of the various marketed fruits, unestimated but rather important quantities of the main export tree crops such as cinnamon, clove, nutmeg, coffee and candle nut (de Foresta and Michon 1997). In Sumatra alone, about 4 million hectares have been converted by local people into various kinds of agroforests without any outside assistance (de Foresta and Michon 1993). An estimated 7 million people in Sumatra and Kalimantan are living from rubber-based agroforests that are spread on about 2.5 million hectares. These agroforests have been established by shifting cultivators through a successional process where tree seedlings are directly planted in the swiddens. The management of the establishment phase constitutes a complex process of forest reconstruction that can be illustrated by the example of damar gardens in Krui, Sumatra (Torquebiau 1984; Michon and Bompard 1987; de Foresta and Michon 1993; Michon *et al.* 1995).

These agroforests start as a classic 'taungya system'. Damar (*Shorea javanica*, a tropical hardwood) seedlings, usually raised in nurseries, are introduced in an already planted rice swidden or most often in a young coffee or pepper plantation established after rice production. This coffee–damar association is maintained up to 10–15 years. It allows seedlings to grow in the best possible conditions in terms of micro-environment and concurrence. However, the parallel with more conventional tree plantations does not go further, and the consecutive phases are more conceived in a logic of association with the forest ecosystem than of environmental confrontation. Once the crop phase is completed, damar trees freely develop with the natural pioneer vegetation that establishes spontaneously. During this period of relative abandonment, through natural processes of dispersion and niche colonisation, the young agroforest gradually acquires a *facies* typical of any secondary forest. This successional forest-garden increases in complexity over years due to a combination between free functioning (development of natural silvigenetic processes) and integrated management (selective cutting and enrichment planting). This management pattern does not fundamentally change when the damar garden starts producing.

In the mature plantation, the balanced combination between natural dynamics and appropriate management of individual trees helps maintain a system which produces and reproduces without disruption in structural or functional patterns. It also allows further diversification through the establishment of more climactic forest species among the cultivated stand. Once established, damar gardens usually reproduce without any major disruption, as decaying trees are replaced whenever needed. Unlike plantations, that evolve through cycles of plantation/ total harvest, damar gardens remain permanent, without reverting to a phase of massive regeneration.

After 40–50 years, the damar plantation reaches its full production period. From a socio-economic point of view, it is not fundamentally different to any specialised commercial plantation: it provides the majority of household income and constitutes an essential complement to ricefields in the farming system (Mary and Michon 1987; Levang

1992). However, from a biological point of view, the mature phase finally resembles more the forest it replaced than a conventional tree plantation. Like a natural forest, it is characterised by a high canopy, a dense undergrowth, relatively high levels of biodiversity and perennial structures. Apart from damar trees that form the frame of the garden, the damar plantation shelters several dozen commonly managed tree species. It is also made up of several hundreds additional species of trees, treelets, liana and epiphytes, spontaneously established and often used. As natural lowland and hill dipterocarp forests in the area are severely depleted, the damar gardens constitute the major habitat for many forest plants and animals, among which are some of the highly endangered mammal species (Sibuea and Herdimansyah 1993; Michon and de Foresta 1995; Thiollay 1995) among others: Sumatran rhinoceros, Sumatran goat, tigers, tapir, gibbons and siamangs.

The 'agroforestry nature' of the Krui damar gardens is more visible in the establishment phase, which constitutes an obvious taungya system associating tree seedlings and annual crops, than in the mature phase that can be analysed as a forest. However, it is the mature phase that really stands where forest and agriculture intersect. Damar gardens have the ecological integrity of a forest, but rely on agricultural practices and are managed mainly as an agricultural enterprise in the middle of farmlands. They perfectly epitomise what integration of forest into farming systems can look like. This addresses the very heart of 'agroforestry'.

Research on all major aspects of damar agroforests have been carried out by a large number of institutions over the last 15 years.[12] These researches have contributed to create an impressive body of solid scientific information that demonstrates the importance of the damar agroforest system as an environmentally sustainable and economically profitable model for the management of both agricultural and forest resources. With little doubt, the process that led to the recent official recognition of the Krui agroforests, benefited from this enormous research effort on damar agroforests. It played an essential role in driving, for the first time ever, the Department of Forestry towards granting legal recognition of farmers rights inside the state forest zone to implement and develop their own management system.

Notes

1. While shifts in government policy towards greater openness to community forestry often coincide with the near depletion of the natural production forest, it may be a simplification to claim, as some do, that communities

receive only what is left over from large-scale commercial exploitation. The shift in the Philippines was, to large extent, due to an overall shift in government policies, in the post-Marcos period, towards local empowerment and poverty alleviation. In Indonesia, greater attention to the welfare of local people in the forest zone has come about, in part, from international pressure as well as from government policies to alleviate poverty in the poorest areas of Indonesia, most of which are found in the Forest Zone.

02. It is interesting to note that officially, many of the university researchers are government officials. But sharing many of the views and approaches of the non-governmental actors, they have been in a good position to play a bridging role between the government and non-government organisations.

03. Indonesia Production Forestry: Achieving Sustainability and Competitiveness, a unpublished working paper on the Indonesian Forestry Sector, 1993 pp. 4–5.

04. Paper is on file with the authors.

05. In 1995, in one operation alone, 450 families in one village were forced from their farms in Lampung Province and elephants were used to uproot more than 4,000 ha of coffee trees. *Lampung Post*, 2 September, 1995 p. 3.

06. The recognition of the Krui agroforestry systems in Lampung, Sumatra and negotiations between government and local communities over village lands in the State Forest Zone in Bentian East Kalimantan, Yamdema and Siberut are examples of this.

07. Members of this informal group include, the University of Indonesia, Latin (a Indonesian forestry and conservation NGO), Watala (a Lampung-based community development NGO), ORSTOM (a French research institute) ICRAF, and the Center for International Forestry Research (CIFOR) Work of this consortium has been supported by a grant from the Ford Foundation.

08. This is a conservative figure based on an average of 200 trees planted in an 15,000 ha area of mature damar agroforests with the State Forest Zone.

09. Surat Keputusan Department of Agraria #5, June 1999.

10. President Suharto resignation came five months after the signing of the KdTI decree.

11. See Van Est and Persoon, this volume, chapter 3 for further discussion of the Philippine CADC experience.

12. In chronological order, the University of Montpellier, France; the National Agronomy Institute of Montpellier, France; SEAMEO-BIOTROP, Indonesia; ORSTOM, France; the University of Toulouse, France; CNRS, France; Himbio, Bandung, Indonesia; the University of Indonesia; the National Center for Agriculture Studies in the Tropics, CNEARC, France; CIRAD-Forêts, France; the Tropical Nature Foundation of Indonesia, LATIN; the Family of Nature and Environment Lovers, WATALA-Lampung, Indonesia; ICRAF; VOCA, USA; CIFOR; the Department of Forestry of Indonesia, LITBANG Kehutanan; the Bogor University of Agriculture, IPB; the University of Paris Sorbonne, France; the University of Orleans, France.

12
Analysing Failed Participation from the Perspective of New Institutional Economics: Evidence from Kerala

V. Santhakumar

1. Introduction

Participation and similar approaches such as collective or community mechanisms have received considerable attention from researchers during the last two to three decades. A major part of this literature explains the factors, or the conditions for the existence, non-existence or the disappearance or sustainability of community based approaches in natural resource management (see, for example, Hardin 1968; Runge 1992; Libecap 1995; Ostrom 1990). There are also studies that analyse the impact of participatory efforts promoted by external agencies (Narayan 1995). This chapter deals with participation promoted by external agencies such as governments and non-governmental organisations. However the central concern here is with reasons for the selection of participation, or a particular mode of participation, by the external agency as a means of achieving its objective. The chapter discusses the conditions under which an external agency promotes a particular mode of participation irrespective of it being the most suitable means of achieving the stated objective. By considering the selection of participation as an issue of institutional choice, the chapter draws on insights from the New Institutional Economics (NIE) (see North 1990; Drobak and Nye 1997). The empirical evidence is drawn from a few cases from the state of Kerala in south India.

2. Analytical framework

In this chapter, the selection of participatory approaches is treated as an issue of institutional choice. To this extent, the insights provided by the New Institutional Economics (NIE) are found to be extremely useful. NIE has attempted to explain institutional changes based on an extended neoclassical framework that allows for boundedly rational behaviour on the part of individuals (North 1990; Drobak and Nye 1997). Non-market institutions are analysed in terms of transactions costs, information asymmetry and uncertainty. Attempts have also been made to explain the persistence of existing institutions within a NIE framework (Arthur 1989; David 1985; North 1990).

The framework assumes that individuals respond to the incentives/disincentives that emerge from the physical, technical, economic and institutional environment. These incentives are not limited to economic benefits and losses, and include non-economic rewards related to prestige, beliefs and ideology. However, economic rationality may play an indirect role in the case of these non-economic rewards. For example, the prevalence of ideology-based decision-making can be seen more in cases where the economic loss associated with such decisions is low (Nelson and Silberberg 1987). Since public actors do not have to bear the cost of policy actions directly, such a framework suggests that their decisions may be more influenced by ideological factors.

These assumptions lead to the following propositions on the effectiveness of an institutional form. An institution can either evolve from within or be 'imposed' by external agents. An evolved institution may be compatible with the material and economic incentives of the situation. However, it is also possible to observe old institutional arrangements that are incapable of further evolution, and unable to respond to changes in the technical and economic environment. In the case of 'imposed' institutions, which are the main concern of this paper, 'success' depends on their compatibility with the material and economic incentives of the situation. If an external agency persists with the promotion of an institutional form, even after it is found to be a failure in meeting stated objectives, this may be due to one or more of the following reasons: the agency does not see it as a failure; or, the agency attributes the non-achievement of the objectives to the poor implementation of the institutional change; or, the actors of the agency have objectives other than those stated.

The first two of these reasons may arise as a result of a number of different processes. There may be lack of information, although this may also be due to the strategic behaviour of some of the actors of the agency. Even if information is available, this may be filtered through ideological or belief-based reasoning, so that failure is not attributed directly to the proposed institutional form. As a result, the agency may not get proper feedback on its own actions. Since rewards to the actors of the agency are not necessarily related to the achievement of stated objectives, this may reduce the pressure on the agency to acquire correct information, or to avoid any form of reasoning that produces an incorrect model of reality. Moreover, at least some actors of the external agency may have a stake in the continuation of the proposed institutional change, irrespective of its success. On the other hand, if actors in the same agency do not have a strong stake in an alternative institutional arrangement, this may lead to a persistent (passive) preference for the selected institutional form, and perhaps the neglect of more efficient ones that could have been deployed in the same situation.

3. Empirical evidence from Kerala

The new institutional framework is used in this chapter to explain some failed attempts to promote participation in the South Indian state of Kerala. The selection of Kerala, and the specific cases within the state, has broader relevance. Kerala is known for its vigorous pursuit of a state-mediated process of development (Dreze and Sen 1993; Isaac and Tharakan 1995). Though the country as a whole accepted this paradigm, Kerala has been seen to be in the vanguard of this process. This occurred through 'public action' in which state interventions were shaped by the 'adversarial' and 'collaborative' participation of social and political groups representing different sections of society (Dreze and Sen 1993; also see Geiser, this volume, chapter 2). These processes are acknowledged to be responsible for the achievement of higher levels of social development in Kerala despite its relatively poor economic record.

It is also generally believed that the strategies and achievements of state interventions (and public action) were not helpful, and possibly counterproductive, in terms of overcoming obstacles for sustained economic growth in the productive sectors of the economy (Isaac and Tharakan 1995). These relatively unsuccessful interventions include efforts to foster institutional changes by the state and related agencies. Promoting participation is one mode of institutional change that was attempted by these agencies. This chapter analyses three such cases, where participation was promoted either by the state or other agencies in the productive sectors of the economy.

In two of the cases described here, group farming and community irrigation, the government promoted participation. This may suggest that difficulties arose due to the general 'insensitivity' of centralised governments to the requirements of specific localities and groups of people. There is a general view among practitioners and scholars of participatory development that the success of participation depends on factors such as participatory learning and design of interventions, efforts to break the inimical influence of local power structures, and the closeness of the organisation promoting participation with the problem situation (Nelson and Wright 1995). In order to examine the impact of organisational characteristics on outcomes, the third case analyses a participatory programme implemented by a well-known non-governmental organisation in Kerala.

3.1 Case I: Group farming

Concern about the decline of paddy cultivation in the state encouraged the Government of Kerala to promote a 'group farming' exercise in 1986. Under this exercise, farmers in each locality were encouraged to join together in the management of agricultural operations. Ownership was retained with the individual farmer. Committees of landowners and government officials were organised, and were responsible for collective farming operations, with one committee covering a land area of 10–50 ha. Government officials, financial institutions, fertiliser agencies, and local panchayats were involved in the committees. The leadership was democratically selected. Committees had considerable flexibility in deciding local technological parameters. The main activities undertaken by committees were soil testing; provision of retail outlets for fertiliser/pesticides or joint transport of these materials; local water management; introduction of mechanisation; and consolidation of other farm operations. The specific incentives were subsidies for manure, pesticides, soil ameliorants and certain machines, as well as support for land development and the establishment of community nurseries. It was presumed that group action would help the organisation of local water management, the use of mechanisation, increase access to credit facilities and reduce the cost of pest management and fertiliser application, due to scale economies.

Over a decade after its introduction, the programme has not helped in reducing the declining trend of paddy cultivation in Kerala. The area under paddy has declined from 0.65 million ha in 1986 to around 0.5 million ha in the mid-1990s. Two micro-level studies, that have assessed the impact or performance of group farming in different villages (Jose 1991; Jacob 1996a), show that most of the participatory activities intended in the programme were not actually carried out. Jacob (1996a) found some cooperative effort in only one of 20 clusters of paddy fields. Instead of joining together, the services of contractors and other hired agents were being used by individual farmers for tilling and other activities. Such private arrangements were also used for obtaining seeds and seedlings. The institutional status of irrigation did not change in any of the villages after the commencement of group farming. Joint efforts for pesticide application were observed only in a few clusters. Only about 10 per cent of farmers used the credit available through the process of group farming, and that too was found only in a few villages. It was observed that group action was primarily undertaken for getting some of the subsidies provided as part of group farming, and that this action has declined as the subsidies have been withdrawn. It seems clear that group action was not sought in most activities and by most of the farmers.

It is well known that some physical, technical, and economic factors reduce the relative profitability of paddy cultivation in Kerala (Oommen 1963; Francis 1990; Krishnan 1991; Narayana 1992; Santhakumar and Rajagopalan 1995). The increasing cost of labour, the increasing real or expected opportunity cost of family labour, climatic factors that reduce paddy productivity, the scope for producing other profitable cash crops which require less labour, and so on, are seen to be major reasons for the reduction of paddy cultivation in the state. The influence of these factors could not be reduced through scale economies under group farming. The expectation that the farmers would use their own or family labour more intensively as part of the group farming exercises also did not materialise. The promoted participatory institution was incapable of reducing the influence of the other factors (i.e. physical, technical and economic factors) that were responsible for the decline of paddy cultivation in Kerala. It may be argued that the failure was due to the poor implementation of group farming, and not due to its conceptual weaknesses. However, there was no demand by farmers for better implementation, which suggests that group farming was not seen to be especially helpful in enhancing profitability. Counterarguments based on the ignorance or inability of farmers to cooperate do not seem to be relevant, since Kerala is well known for participatory efforts (even by farmers) to demand and acquire public resources for social services such as health and education.

3.2   Case II: Participation in irrigation management

Observing that the irrigation bureaucracy had failed to carry out field-level water management effectively, the Government of India initiated the formation of farmers' committees to manage field-level irrigation from the 1970s (Pant 1999). These committees, organised under the umbrella of 'Command Area Development Authority', were expected initially to build up infrastructure for field-level water delivery, and later to take over water management on their own. Such committees were started in the irrigation command areas of Kerala too. The committees were to collaborate with the irrigation department in the construction of field channels, farm roads, and distribution channels. They were also responsible for maintaining the canals and channels, and for field-level water management. Subsidies were available for the construction of field and distribution channels, as well as a management subsidy based on area covered.

However, the experience of this form of promoted participation in irrigation management in Kerala has not been encouraging. Most of the farmer's committees formed for managing field-level irrigation have become inactive (Jacob 1996). The farmers have not even surrendered the required land for the construction of field channels. The subsidies have not been used by a large number of farmers. Such indifference to the participatory mechanism does not mean that farmers are neither using irrigation nor involved in any group action for getting water. People depend on other arrangements, including self-initiated collective action for irrigation in localities where farmer's committees formed by the government are inactive. Table 12.1 provides a list of some of these alternative arrangements, as observed in the command area of an irrigation project (Neetha, forthcoming). The data suggest that more than 98 per cent of farmers are indifferent to participatory action promoted by the government. Over 83 per cent either make individual investments, or buy water from a merchant, or are involved in group actions of a different kind.

Two key factors seem to have encouraged the government to promote participatory action. The first was an assumption that the same type of group action would be suitable for all the farmers living in all the command areas of India. Though the government was aware of the limitations of canal irrigation, its planners felt that these problems could be solved through the implementation of a command area development programme through farmers' committees. The reality, as is evident from the response of farmers, was that each farmer faced a specific set of problems in getting water, and had a specific set of resources at his command. For example, a farmer whose plot is at a higher elevation and is also employed in a non-agricultural occupation would have a different level of problems and resources compared to another farmer

Table 12.1   Different arrangements existing for irrigation

Arrangement	Percentage of irrigated plots	Percentage of irrigated area
No action (total dependence on public provision)	14.6	13.6
Participation initiated by the Government	1.5	1.7
Different types of group action by the people themselves	25.2	23.7
Buying water	3.3	2.9
Individual investment such as digging a well	43.3	46.6
Buying water and individual investment	9.3	9.7
Individual investment and group action	2.8	1.8

Source: Neetha (forthcoming).

whose plot is a marshy field and who is employed in agriculture on a full-time basis. Material conditions, such as the position of plots, and economic features, such as the opportunity cost of time of the individual farmer, would influence the selection of arrangements, and importantly, decisions about whether to participate in the group action promoted by the government. The second assumption was that farmers were incapable of taking part in group action without the encouragement of the government. However, the existence of different arrangements made by the people suggests that farmers are able to organise group action and to undertake individual or private transactions to meet their objective of getting water.

3.3 Case III: Participatory planning for local development

The Kerala Sastra Sahithya Parishat (KSSP) is a well-known movement involved in the popularisation of science, environmental protection and decentralised development (Govindan 1989; Guha 1988).[1] KSSP on its own, and through its linkages with other socio-political movements, has grassroot connections all over the state. KSSP has been advocating decentralised governance, and building local capacity for self-governance. Its objective is to promote a participatory democratic process aimed at local-level development. It facilitates participation from the beginning of the development process, through the participatory mapping of local resources and the preparation of development plans. KSSP sought the involvement of local NGOs, popular movements including political parties, and the active participation of the elected representatives of local governments in the preparation of development plans. These plans were expected to contribute towards informed local governance and the development of rural areas of the state. The activities proposed in the local plans included increasing agricultural productivity, starting small-scale industries, and the creation of physical and social infrastructure.

Given that this form of participatory planning was carried out by a popular movement in a state which is known for high levels of social development, and for its success in breaking traditional power structures, one would expect it to have overcome the factors that limited the success of state-led participatory development. However, there are indications that the implementation of these local-level plans has turned out to be extremely difficult.[2] The local people are reluctant to take up the suggested developmental activities, especially those that aim at the economic development of the villages in sectors such as agriculture and industry.

KSSP assumed that it was the lack of knowledge of resources or problems that obstructed rural development. However, there is evidence that people acquire, and are capable of acquiring information, from multiple sources if activities are seen to economically beneficial. This can be seen from the rapid spread of rubber cultivation in the state, and widespread migration to the Middle East in search for employment. Further, it was thought that unemployment was due to lack of job opportunities in rural areas. While this is true to some extent, there is a preference for jobs in government and the formal sector, which is facilitated through the spread of free education and integration with the national and international labour market. In fact there is a severe shortage of agricultural labour in the villages (Francis 1990), and the increase of such job opportunities may not attract unemployed people. It was also thought that the development of productive activities within the region was necessary. Although this is important in the long term, in the short term, people may opt for outside employment if it is more lucrative than working locally or on their own land. Finally, there was a belief that group action could accelerate rural development. However, the slow pace of investment in rural areas is due to the high (perceived and real) cost of labour, land, electricity and other resources. Lack of capital is not a major constraint, as is evident from the low credit-deposit ratio in the banks of Kerala. These hurdles do not seem to be ones that can be usefully avoided through group action.

4. Analysing institutional choice and persistence

Having seen that these participatory initiatives were unsuccessful, the following questions are relevant to the analysis of institutional choice and persistence: (1) why did external agencies adopt participatory approaches?; and (2) why did they persist with this approach, in spite of signals of their failure?

4.1   Group farming

The government of Kerala was functioning within the framework of state-mediated development that was adopted and followed by the governments of independent India and its different states during the last four decades. Under this framework, subsidies were given to farmers for different inputs for crop production and for enhancing productivity (Vaidyanathan 1996). Since beneficiaries did not directly bear any costs in such schemes, they had little incentive to demand effectiveness or efficiency in state intervention. Thus, public decision-makers could be relatively unconcerned about the actual requirements of the people in the formulation and implementation of such programmes. This chapter suggests that such a state-mediated process of development can be seen to provide public decision-makers with *incentives to neglect the incentives* of the targeted beneficiaries.

Successive governments of Kerala have been eager to implement programmes formulated within this framework of state-mediated development. The leftist political parties of Kerala and other parties, which respond to the demands of organised trade unions and pressure groups when they are in power, have a record of 'sincerely' implementing programmes of public investment for social development and welfare schemes, as well as land reforms (Isaac and Tharakan 1995). In addition to this more loyal attitude of Kerala politicians to state-mediated development, other factors attracted them to the nationally conceived programmes for food crop production. These subsidy programmes suited their own efforts to enhance paddy production within the state, due to their declared intention of achieving food self-sufficiency (Santhakumar *et al.* 1995; Santhakumar and Rajagopalan 1995). The need for more agricultural labour for paddy production compared to other crops also encouraged the left political parties, which were influenced by agricultural labour unions, to embrace the theme of rice self-sufficiency.

These perceptions encouraged political decision-makers in Kerala to vigorously pursue a process of state-mediated development in agriculture. Under such conditions, there was little pressure to collect information to assess the effectiveness of proposed interventions, and an increased likelihood of taking decisions based on inadequate information, or on an incorrect assimilation of available information. For instance, the influence of agro-climatic and technical factors, which result in the low profitability of paddy cultivation in Kerala, was well known by the 1980s (Chandler 1979; Gangadharan 1985). However, planners in Kerala were either ignorant or reluctant to accept the limiting role of climatic factors in their efforts to increase paddy productivity even in the late 1980s (Santhakumar and Rajagopalan 1995), when they formulated the group farming programme to boost paddy production. Furthermore, the role of economic factors, such as the possibility of cultivating more profitable crops, increase in wage rates, the disinterest of part-time farmers, and so on were highlighted in the social discourses of Kerala during the last couple of decades (NCAER 1962; Oommen 1963). At one level, planners thought that such disincentives for paddy cultivation could be overcome partly through the subsidies provided, and also through the potential benefits of the group action. Planners also perceived the need to influence farmers' behaviour, by creating awareness about the supposed need for achieving self-sufficiency in rice production within Kerala. This perception, together with a view of farmers as ignorant and malleable, reinforced the belief that farmers' responses could be modified through a combination of subsidies and information.

Convinced of the need, and the possibility of compelling farmers, to cultivate paddy, planners looked for strategies to influence farmers' behaviour. Group action got precedence, not due to its proven superiority to achieve the stated objective, but because it fitted well with the preference of the left political groups for cooperative and collective action. Due to the failure of collective farms both within Kerala, and in other parts of the world, the left political groups which ruled Kerala in 1986 did not want to reattempt collective exercises that precluded the private ownership of land by farmers. Thus, group farming, which aimed at combining collective farm operations and private property rights, was an attractive proposition to the political decision-makers of Kerala.

While this may be a reasonable explanation for the adoption of group farming by the Government of Kerala, understanding the persistence of this approach requires an analysis of the feedback process. In spite of signals of its failure, the Government of Kerala does not appear ready to abandon the idea of group action for paddy cultivation. Political decision-makers can be expected to respond to the feedback provided by actors such as farmers (direct beneficiaries), the bureaucracy (implementing arm of the state), and the public (general taxpayers). However, farmers have little incentive to abandon the programme since they do not bear its direct cost. Whatever they get is a net gain, and they are not concerned about the effectiveness or efficiency of the selected approach. The bureaucracy has incentives for the continuation of the approach, since it gives them more positions and opportunities for the provision of subsidies, creating awareness, monitoring and so on. The failures in this regard are not costly to them, given the existing structure of accountability. Thus, the benefits of alternative institutional approaches are unknown and uncertain to both the farmers and bureaucracy, and their feedback can be expected to be either status-quoist or indifferent. The public in general probably also have few incentives to demand a reformulation of state-mediated processes such as group farming, because the number of direct tax payers are a minority, while a majority of people enjoy some form of subsidy. Under such conditions, only self-reinforcing feedback is generated, and political decision-makers are happy to continue with such approaches, that satisfy their ideological moorings, as well as giving them popular slogans such as self-sufficiency. Even for these decision-makers, the political benefits of more efficient state investment are uncertain. Thus, the influence of factors that resulted in the selection of group farming has not been countered by negative feedback due to actual experience, and this can be seen to explain its persistence.

4.2   Participation in irrigation management

The assumptions behind the formulation of a national programme for community participation in irrigation management were not appropriate to the physical and economic conditions faced by farmers in Kerala. The physical conditions included the topography, rainfall availability, and the cropping pattern. The economic conditions included the relative prices of crops, cost of wage labour, opportunity cost of family labour and so on. Thus, it seems evident that the failure of community participation in irrigation was also due to the incompatibility of the promoted institutional structure with the incentives of the physical and economic environment.

The reasons that led to the adoption of this inappropriate institutional form are not difficult to trace. The Command Area Development

(CAD) programme was a nationally conceived activity, and its approach was oriented to the problems of the major states of the country. Under such a national programme, it is easy to neglect the specific topographical, climatic and economic issues of a small wet-tropical state such as Kerala. However, once a national programme and an allocation of funds exist, the political and the technical planners of every individual state have incentives to try and get a share of this allocation, and to implement the programme in their own state, irrespective of whether it is appropriate to local conditions. Since non-implementation would deprive the state of financial resources allocated for this purpose, it is rational on the part of every state to follow the national pattern. Furthermore, there were strong incentives for the technical bureaucracy to adopt such a programme, since the kind of participation envisaged under the nationally conceived programmes provided them with more positions and opportunities compared to the self-initiated arrangements of the irrigators. Thus, the nationally conceived CAD programme provided incentives for every region (including Kerala) to adopt it, irrespective of whether it was appropriate to local conditions.

The reasons for the persistence of this approach, despite signals that it was inappropriate or ineffective, can also be understood by analysing the feedback provided by different actors. As in the case of group farming, the irrigators have little incentive to demand the abandonment of CAD programme, even if it does not help them get water more effectively and efficiently. The positions and opportunities available through CADA to the Irrigation Department of Kerala state do not encourage the bureaucracy to attempt to change the status quo, especially when their benefits under alternative arrangements are uncertain. One cannot expect any response from the general public, when only a small group pay direct tax and few make a connection between the inefficiency of a specific programme and taxes, or the efficient delivery of public goods in general. The government of Kerala does not have incentives to alter the programme, since they do not directly bear the cost, and because of the self-reinforcing feedback provided by the Irrigation Department and the indifferent attitude of irrigators and the general public. This results in the persistence of this specific mode of participatory activity for irrigation management in Kerala.

4.3 Participatory planning for local development

The assumptions made by KSSP indicate that the organisation believed in a planned interventionist paradigm of development, and considered local people as ignorant and malleable. KSSP's tradition was of a neo-Marxist popular movement (Guha 1988, Govindan 1989), and its perception of participation was shaped by its belief in decentralised but planned economic development and the importance of inculcating a scientific attitude among the so-called 'ignorant' people. KSSP attempted to initiate a participatory planning process in the context of this ideological mindset. The availability of funds at the national and state levels, as well as the discourses of participatory development dominant in the 1980s, facilitated this strategy.

KSSP found it relatively easy to mobilise a large number of volunteers for short-term programmes like planning, but this voluntary spirit was not sustained for the implementation of developmental activities, which require longer-term commitments. This successful participation in planning activities should have given the organisation adequate indications about the untenability of their own assumptions, and the factors that might make the implementation of rural development plans difficult. However this has not happened. There can be two reasons for this: first, it is very difficult to assimilate information contrary to one's own beliefs, especially if the cost of such selective assimilation is not high, as is the case in a public organisation such as KSSP. Second, the feedback provided, or the preferences revealed, by the local people in the participatory planning exercises, also served to reinforce the beliefs of the external agency, rather than contradicting them. This was because the local people recognised that funds for implementing the proposed developmental activities would eventually come from governmental or other external sources, and that they would not have to bear the costs either directly or indirectly. The NGO was not ready to assess the willingness of the local people to bear the cost of suggested activities, because of its own commitment to state-mediated development. This perception of 'benefits at no cost' made the local beneficiaries demand actions without considering the opportunity costs of resources, and also to provide positive responses to the suggestions that emanated from the external agency promoting participatory development. This reassuring (but unrealistic) feedback can be seen to have limited the capacity of the participatory planning process to generate information on the factors that constrain local-level development in Kerala.

5. Conclusions

The discussion of these cases from Kerala does not imply that group farming, community irrigation or participatory planning would fail in all circumstances. Such strategies can be effective in several situations. For instance, in the case of community irrigation, documented evidence exists suggesting its effectiveness in some Indian states (Pant 1999). This chapter has argued that these participatory efforts have failed in Kerala due to their incompatibility with physical, technical and economic factors in specific situations. This argument does not imply that participation can never succeed in Kerala. There is evidence of the collaborative and adversarial participation by the people of Kerala in the context of state interventions in sectors such as health and education (see also Geiser, this volume, chapter 2).

Why is it that targeted beneficiaries remain mute spectators to a programme, even if it is not beneficial to themselves? It is here that the issue of appropriate feedback becomes important. If a strategy is such that the targeted beneficiaries do not lose anything, even if the strategy is ineffective, there is little incentive for the beneficiaries to demand a change in strategy. Moreover, even if the strategy fails to achieve the stated objectives, there will be non-targeted benefits for some beneficiaries. These include the prospects for diverting subsidies, leadership incentives for some people, political mileage for local organisations, and so on. Such non-targeted benefits further reduce the prospects of a strong demand for a change in strategy. In all the cases of participation described here, the cost of institutional change was borne by the external agency, and beneficiaries did not provide any negative feedback to those promoting the new strategies.

It can be argued that external interventions should not be based on preconceived judgements, and should be flexible enough to respond to the specific requirements of the locality. However, a modification of the assumptions that seemed to have influenced decision-making in Kerala would require significant changes in the mind-set of planners, as well as in the incentive structure of public agencies. Given the role of public agencies, and the persistence of their incentive structure, it is difficult to envisage such a radical change. The structural factors that shape and sustain the traditional approach cannot be countered, in my view, through the use of methodologies such as participatory appraisals. I would argue that the political economy of participation, in which external funding agencies, national and state governments, and non-governmental organisations attempt to institute organisational changes in supposedly 'traditional' and 'backward' societies, provides sufficient incentives for the continued implementation of participatory programmes that are based on a poor understanding of what is institutionally appropriate in a local context.

Notes

1. The analysis is based on the experience of the author as a full-time volunteer and functionary of the NGO, for a period of eight years. He was closely associated with the process of micro-watershed management (as coordinator) and the mapping programmes.
2. This is evident from the experience of Kalliassery panchayat of North Kerala, where the NGO started the programme in 1990, and where a large number of initiatives have been taken to start the implementation of the plan. However, they have been unable to implement any major project in the productive sectors of the rural economy.

References

Abbot, J. and I. Guijt. 1998. *Changing Views on Change: Participatory Approaches to Monitoring the Environment*. SARL Discussion Paper 2, London: International Institute for Environment and Development.

Adam, B. 1998. *Time Scapes of Modernity: The Environment and Invisible Hazards*. London: Routledge.

Adnan, S., A. Barrett, S. M. Nurul Alam and A. Brustinow. 1992. *People's Participation, NGOs and the Flood Action Plan – An Independent Review*. Commissioned by Oxfam-Bangladesh. Dhaka: Research & Advisory Services.

Agarwal, B. 1997. Environmental Action, Gender Equity and Women's Participation. *Development and Change* 28: 1–44.

Agrawal, A. and C. C. Gibson. 1999. Enchantment and Disenchantment: The Role of Community in Natural Resource Conservation. *World Development* 27, 4: 629–49.

Anderson, J., J. Clement and L. V. Crowder. 1998. Accommodating Conflicting Interests in Forestry: Concepts Emerging from Pluralism. *Unaslyva* 49, 194: 3–10.

Arthur, W. B. 1989. Competing Technologies, Increasing Returns, and Lock-in by Historical Events. *Economic Journal* 99: 116–31.

Attwood, D. W. and B. S. Baviskar. 1988. *Who Shares? Cooperatives and Rural Development*. New Delhi: Oxford University Press.

Bahuguna, V. K. 1992. *Collective Resource Management: An Experience in Harda Forest Division*. Bhopal: Regional Centre for Wastelands Development, Indian Institute of Forest Management.

Bahuguna, V. K. and V. Luthra. 1991. *Forest Administration in India: Policy, Institutional and Organisational Issues*. Bhopal: Indian Institute of Forest Management.

Baland, J. M. and J. P. Platteau. 1996. *Halting Degradation of Natural Resources: Is there a Role for Rural Communities?* Oxford: Clarendon Press.

Barraclough, S. L. and K. B. Ghimire. 1995. *Forests and Livelihoods: The Social Dynamics of Deforestation in Developing Countries*. New York: Macmillan – now Palgrave.

Bates, R. 1988. Contra Contractarianism: Some Reflections on the New Institutionalism. *Politics and Society* 16: 387–401.

Bates, R. 1995. Social Dilemmas and Rational Individuals: An Assessment of the New Institutionalism pp. 27–48. In *The New Institutional Economics and Third World Development* (eds) J. Harriss, J. Hunter and C. M. Lewis. London: Routledge.

Battaglia, D. 1995. On Practical Nostalgia: Self-Prospecting among Urban Trobrianders, pp. 77–96. In *Rhetoric's of Self-Making* (ed.) D. Battaglia. Berkeley: University of California Press.

Baumol, W. J. and W. E. Oates. 1988. *The Theory of Environmental Policy*. Cambridge: Cambridge University Press.

Baviskar, A. 1999. Participating in Ecodevelopment: The Case of the Great Himalayan National Park, Himachal Pradesh. In *A New Moral Economy for India's Forests?* (eds) R. Jeffery and N. Sundar. New Delhi: Sage.

Beck, T. 1994. *The Experience of Poverty: Fighting for Respect and Resources in Village India.* London: Intermediate Technology Publications.

Benda-Beckmann, F. von and A. Brouwer. 1992. *Changing Indigenous Environmental Law in the Central Moluccas: Communal Relations and Privatisation of Sasi.* Paper presented to the Congress of the Commission on Folk Law and Legal Pluralism at Victoria University, Wellington.

Bennagen, P. L. and M. L. Lucas-Fernan (eds), 1996. *Consulting the Spirits, Working with Nature, Sharing with Others: Indigenous Resource Use in the Philippines.* Quezon City: Sentro Para sa Ganap na Pamayanan.

Berkes, F. (ed.) 1990. *Common Property Resources: Ecology and Community-based Sustainable Development.* London: Belhaven Press.

Bernard, H. R. 1994. *Research Methods in Anthropology. Qualitative and Quantitative Approach.* London: Sage Publications.

Bierschenk, T. 1988. Development Projects as Arenas of Negotiation for Strategic Groups: A Case Study from Benin. *Sociologica Ruralis* **28**, 2/3: 146–60.

Blaikie, P. 1997. Classics in Human Geography Revisited: Author's Response. *Progress in Human Geography* **21**, 1: 79–80.

Blauert, J. and E. Quintanar. 1997. Seeking Local Indicators: Participatory Stakeholder Evaluation of Farmer-to-Farmer Projects (Mexico). Paper presented at the International Workshop on Participatory Monitoring and Evaluation: Experiences and Lessons. November 24–29, 1997. IIRR Campus, Silang, Cavite, Philippines.

Boissevain, J. 1992. On Predicting the Future: Parish Rituals and Patronage in Malta. In *Contemporary Futures. Perspectives from Social Anthropology* (ed.) S. Walmann. London: Routledge.

Borrini-Feyerabend, G. 1996. *Collaborative Management of Protected Areas: Tailoring the Approach to the Context.* Gland: IUCN, The World Conservation Union.

Borrini-Feyerabend, G. 1997. *Beyond Fences: Seeking Social Sustainability in Conservation.* Gland: IUCN, The World Conservation Union.

Bourdieu, P. 1963. The Attitude of the Algerian Peasant Toward Time, pp. 55–72. In *Mediterranean Countrymen: Essays in the Social Anthropology of the Mediterranean* (ed.) J. Pitt-Rivers. Paris/La Haye: Mouton & Co.

Bromley, D. W. (ed.) 1992. *Making the Commons Work.* San Francisco: ICS Press.

Burgess, N. D. and C. Muir (eds) 1994. Coastal Forests of Eastern Africa: Biodiversity and Conservation. Sandy: Society for Environmental Exploration/Royal Society for the Protection of Birds.

Carney, D. (ed.) 1998. *Sustainable Rural Livelihoods – What Contribution Can We Make?* London: Department for International Development.

Chambers, R. 1991. *Rapid and Participatory Rural Appraisal.* Brighton: Institute of Development Studies, University of Sussex.

Chambers, R. 1983. *Rural Development: Putting the Last First.* New York: Longman.

Chambers, R. 1995. *Poverty and Livelihoods: Whose Reality Counts?*, Discussion Paper 347, Brighton: Institute for Development Studies, University of Sussex.

Chambers, R. 1997. *Whose Reality Counts? Putting the Last First.* London: Intermediate Technology Publications.

Chambers, R. and G. Conway. 1992. *Sustainable Rural Livelihoods: Practical Concepts for the 21st Century*. Discussion Paper 296, Brighton: Institute for Development Studies, University of Sussex.

Chambers, R., A. Pacey and L. A. Thrubb 1989. *Farmer First: Farmer, Innovation and Agricultural Research*. London: Intermediate Technology Publications.

Chambers, R., N. C. Saxena and T. Shah. 1989. *To the Hands of the Poor: Water and Trees*. London: Intermediate Technology Publications.

Chandler, R. F. Jr. 1979. *Rice in the Tropics: A Guide to the Development of National Programs*. Boulder: Westview Press.

Chopra, K., G. K. Kadekodi, and M. N. Murty. 1990. *Participatory Development: People and Common Property Resources*. New Delhi: Sage Publications.

Clark, A. 1995. India: Andhra Pradesh Forestry Project. In *World Bank Sourcebook on Participation*. Washington DC: Environment Department, The World Bank.

Coase, R. 1937. The Nature of the Firm. *Economica* **4**, 3: 386–404.

Coleman, J. 1990. *Foundations of Social Theory*. Cambridge, MA: Belknap.

Colfer, C. J. P., D. W. Gill and A. Fahmuddin. 1988. An Indigenous Agricultural Model from West Sumatra: A Source of Scientific Insight. *Agricultural Systems* **26**: 191–209.

Conelly, W. T. 1992. Agricultural Intensification in a Philippine Frontier Community: Impact on Labour Efficiency and Farm Diversity. *Human Ecology* **20**, 2: 203–21.

Conroy, C. and M. Litvinoff (eds) 1988. *The Greening of Aid: Sustainable Livelihoods in Practice*. London: Earthscan.

Conroy, C., A. Mishra and A. Rai. 1999. *Self-Initiated Community Forest Management in Orissa: Practices, Prospects and Policy Implications*. Chatham, UK: Natural Resources Institute.

Cousins, B. 1995. *A Role for Common Property Institutions in Land Redistribution Programmes in South Africa*. Gatekeeper Series No. 53, London: International Institute for Environment and Development.

CPFD. 1997. *The Community and Private Forestry Programme in Nepal*. Kathmandu: Community and Private Forest Division, Department of Forest.

Crehan, K. 1997. *The Fractured Community: Landscapes of Power and Gender in Rural Zambia*. Berkeley: University of California Press.

D'Silva, E. (ed.), 1995. *The Role of Forest Departments in the 21st Century*. Washington DC, New Delhi and Hyderabad: Economic Development Institute of the World Bank, Ministry of Environment and Forests, and Andhra Pradesh Forest Department.

D'Silva, E. 1997. Why Institutional Reforms in Forestry? Lessons from International Experience. *Natural Resources Forum* **21**, 1: 51–60.

David, P. A. 1985. Clio and the Economics of QWERTY. *American Economic Review* **75**: 332–7.

Davidson, J. and D. Myers, with M. Chakraborty (1992) *No Time to Waste*. Oxford: Oxfam.

de Foresta, H. and G. Michon. 1993. Creation and Management of Rural Agroforests in Indonesia: Potential Applications in Africa, pp. 709–24 in *Tropical Forests, People and Food: Biocultural Interactions and Applications to Development* (eds) C. M. Hladik *et al*. Paris: UNESCO and the Parthenon Publishing Group.

de Foresta, H. and G. Michon 1997. The Agroforest Alternative to *Imperata* Grasslands: When Smallholder Agriculture and Forestry Reach Sustainability. *Agroforestry Systems* **36**: 105–20.

de Groot, W. T. 1992. *Environmental Science Theory: Concepts and Methods in a One-World, Problem Oriented Paradigm*. Amsterdam: Elsevier Science Publishers.

de Jong W. 1994. Recreating the Forest: Successful Examples of Ethno-conservation Among Land-dayaks in Central West Kalimantan. Communication to the *International Symposium on Management of Tropical Forests in Southeast Asia*, Oslo, Norway, March.

Dearden, P., M. Carter, J. Davis, R. Kowalski and M. Surridge. 1999. Icitrap – An Experiential Training Exercise for Examining Participatory Approaches to Project Management. *Public Administration and Development* **19**: 93–104.

Dijk, R. van. 1997. *Fundamentalism, Cultural Memory and the State: Contested Representations of Time in Post-colonial Malawi*. Working Paper 2. Den Haag: WOTRO.

Dobson, A. (ed.) 1999. *Fairness and Futurity*. Oxford: Oxford University Press.

Donnelly, A., B. Dalal-Clayton and R. Hughes. 1998. *A Directory of Impact Assessment Guidelines – Second Edition*. London: International Institute for Environment and Development.

Dove, M. R. 1993. Smallholder Rubber and Swidden Agriculture in Borneo: A Sustainable Adaptation to the Ecology and Economy of the Tropical Forest. *Economic Botany* **47**, 2: 136–47.

Dreze, J and A. Sen. 1993. *Hunger and Public Action*. Delhi: Oxford University Press.

Drijver, C. A. 1990. People's Participation in Environmental Projects in Developing Countries, pp. 528–40. In *The People's Role in Wetland Management. Proceedings of the International Conference Leiden, The Netherlands, June 5–8, 1989* (eds) M. Marchand and H. Udo De Haas. Leiden: Reproductie afdeling Biologie.

Drijver, C. A. 1993. Participatory Rural Appraisal: A Challenge for People and Protected Areas. CML Coursebook III, 1994–1995, Leiden: Centre of Environmental Science.

Drobak, J. N. and J. V. C. Nye (eds) 1997. *The Frontiers of the New Institutional Economics*. San Diego: Academic Press.

Eade, D. and S. Williams. 1995. *Oxfam Handbook for Development and Relief*. Oxford: Oxfam.

Engel, P. G. H., M. L. Salomon and M. E. Fernandez. 1994. *RAAKS, A Participatory Methodology for Improving Performance in Extension, Version 5.1*. Wageningen: WAU/CTA/IAC.

Esman, M. J and N. T. Uphoff. 1984. *Local Organisations: Intermediaries in Rural Development*. Ithaca: Cornell University Press.

Est, D. van and R. Noorduyn. 1997. Management of Natural Resource Conflicts: A Case from the Logone Floodplain in North-Cameroon. Paper presented at the Seminar *Managing the Dry African Savannah: Options for Conservation and Sustainable Use*. Leiden: Leiden University.

Estrella, M. and J. Gaventa. 1997. *Who Counts Reality? Participatory Monitoring and Evaluation: A Literature Review*. Brighton: Institute for Development Studies, University of Sussex.

FAO. 1988. *Participatory Monitoring and Evaluation: Handbook for Training Fieldworkers*. Bangkok: Regional Office for Asia and the Pacific, Food and Agriculture Organisation of the United Nations.

Farrington, J., A. Bebbington and K. Wellard. 1993. *Between the State and the Rural Poor: NGOs and Sustainable Development*. London: Routledge.

Ferrer, E. and C. Nozawa. 1997. Community-based Coastal Resource Management in the Philippines: Key Concepts, Methods and Lessons Learned. Paper presented at the International Development Research Centre Planning Workshop on Community-based Natural Resource Management, 12–16 May 1997, Hué, Vietnam.

Fisher, R. J. 1991. Local Organisations in Community Forestry. Paper presented at the *FAO Regional Wood Energy Development Programme Expert Consultation Local Organisations in Forestry Extension*. Chiang Mai, Thailand, 7–12 October.

Fisher, R. J. 1995. *Collaborative Management of Resources for Conservation and Development*. Issues in Forest Conservation, Gland: IUCN, The World Conservation Union.

Fisher, R., W. Ury and B. Patton. 1992; 2nd edn 1997. *Getting to Yes: Negotiating an Agreement Without Giving In*. London: Arrow Books Ltd.

Francis, S. K. 1990. *Dynamics of Rural Labour Market: An Analysis of Emerging Agricultural Labour Shortage in Kerala Region*. Unpublished M.Phil. thesis, Trivandrum: Centre for Development Studies.

Fukuyama, F. 1995. *Trust: The Social Virtues and the Creation of Prosperity*. New York: Free Press.

Gadgil, M. and R. Guha. 1993. *This Fissured Land: An Ecological History of India*. New Delhi: Oxford India Paperbacks.

Gangadharan, C. 1985. Breeding. in *Rice Research in India* (ed.) P. L. Jaiswal. New Delhi: Indian Council for Agricultural Research.

Gell, A. 1996. *The Anthropology of Time: Cultural Constructions of Temporal Maps and Images*. Oxford: Berg.

Ghai, D. and J. M. Vivian (eds) 1992. *Grassroots Environmental Action: People's Participation in Sustainable Development*. London: Routledge.

Gibson, U and T. Koonz. 1998. When 'Community' is not Enough: Institutions and Values in Community-Based Forest Management in Southern Indiana. *Human Ecology* **26**, 4: 621–47.

Giddens, A. 1984. *The Constitution of Society: An Outline of the Theory of Structuration*. Cambridge: Polity Press.

Gilmour, D. A. 1990. Resource Availability and Indigenous Forest Management Systems. *Society and Natural Resources* 3: 145–58.

Gilmour, D. A. and R. J. Fisher. 1991. *Villagers, Forests and Foresters: The Philosophy, Process and Practice of Community Forestry in Nepal*. Kathmandu: Sahayogi Press.

Godoy, R., N. Brokaw and D. Wilkie. 1995. The Effect of Income on the Extraction of Non-timber Tropical Forest Products: Model, Hypotheses, and Preliminary Findings from the Sumu Indians of Nicaragua. *Human Ecology* 23: 29–52.

Gopalan, A. K. 1973. *In the Cause of the People: Reminiscences*. Bombay: Orient Longman.

Gorman, M. 1995. Report on the Socio-Economic Study/Participatory Rural Appraisal, conducted in March/April 1995. Tanga Coastal Zone Conservation and Development Programme, Tanzania.

Gorman, M., G. Uronu, S. Semtaga, A. Mfuko, J. Kabamba, M. Mfuko, L. Challenge, R. Mosha, M. Kwalloh, A. Mahanyu, and A. Seumbe. 1996. Assessment of Participation and Village Support for Pilot Environmental

Committees. Tanga Coastal Zone Conservation and Development Programme. Internal Report. October–November.

Gouyon A., de Foresta H. and P. Levang. 1993. Does 'Jungle Rubber' Deserve its Name? An Analysis of Rubber Agroforestry Systems in Southeast Sumatra. *Agroforestry Systems* 22: 181–206.

Govindan, P. 1989. Science for Social Revolution: Science and Culture in Kerala. *Impact of Science on Society* 155: 233–40.

Grimble, R. and M.-K. Chan. 1995. Stakeholder Analysis for Natural Resource Management in Developing Countries: Some Practical Guidelines for Making Management More Participatory and Effective. *Natural Resources Forum* 19, 2: 113–24.

Grimble, R., M. and K. Chan, J. Aglionby and J. Quan. 1995. *Trees and Trade-Offs: A Stakeholder Approach to Natural Resource Management.* Gatekeeper Series No. 52. London: International Institute for Environment and Development.

Grimble, R.-K. Wellard. 1997. Stakeholder Methodologies in Natural Resource Management: A Review of Principles, Contexts, Experiences and Opportunities. *Agricultural Systems* 55, 2: 173–93.

Grindle, M. S. and M. E. Hildebrand. 1995. Building Sustainable Capacity in the Public Sector: What can be Done? *Public Administration and Development* 15: 441–63.

Grindle, M. S. and J. W. Thomas. 1991. *Public Choices and Policy Change: The Political Economy of Reform in Developing Countries.* Baltimore and London: The Johns Hopkins University Press.

Grove, R. 1998. The East India Company, the Raj, and El Nino: The Critical Role Played by Colonial Scientists in Establishing the Mechanisms of Teleconnection 1770–1930. In *Nature and the Orient* (eds) R. Grove, V. Damodaran and S. Sangwan. Delhi: Oxford University Press.

Guha, R. 1988. Ideological Trends in Indian Environmentalism. *Economic and Political Weekly* 23, 49: 2578–81.

Guha, R. and J. Martinez-Alier. 1997. *Varieties of Environmentalism: Essays North and South.* London: Earthscan.

Guijt, I. 1996. *Questions of Difference – PRA, Gender and Environment, A Training Video.* London: International Institute for Environment and Development.

Guijt, I. and M. M. K. Shah. 1998. *The Myth of Community: Gender Issues in Participatory Development.* London: Intermediate Technology Publications.

Hadenius, A. and F. Uggla. 1996. Making Civil Society Work, Promoting Democratic Development: What Can States and Donors Do? *World Development* 24, 10: 1621–39.

Haeuber, R. 1993. Development and Deforestation: Indian Forestry in Perspective. *The Journal of Developing Areas* 27: 485–514.

Hardin, G. 1968. The Tragedy of the Commons. *Science* 162: 1243–8.

Harris, M. 1991. *Cultural Anthropology.* New York: HarperCollins.

Hegde, R., S. Suryaprakash, L. Achoth, and K. S. Bawa. 1996. Extraction of Non-timber Forest Products in the Forests of Biligiri Rangan Hills, India: Contribution to Rural Income. *Economic Botany* 50: 243–51.

Hildyard, N., P. Hegde, P. Wolverkamp and S. Reddy. 1998. Same Platform, Different Train: The Politics of Participation. *Unaslyva* 49, 194: 3–10.

Hill, I. and D. Shields. 1998. *Incentives for Joint Forest Management in India. Analytical Methods and Case Studies.* World Bank Technical Paper No. 394. Washington, DC: IBRD.

Hobley, M. 1996. *Participatory Forestry: The Process of Change in India and Nepal.* Rural Development Forestry Study Guide 3. London: Overseas Development Institute.

Hobley, M. and K. Shah. 1996. What Makes a Local Organisation Robust? Evidence from India and Nepal. *Natural Resource Perspectives Number 11,* London: Overseas Development Institute.

Horrill, J. C. and H. Kalombo. n.d. Kipumbwi Reef Fisheries Action Plan: Background and Process. Internal Report. Tanga Coastal Zone Conservation and Development Programme, Tanzania.

Hughes, R. and B. Dalal-Clayton. 1996. *Participation in Environmental Impact Assessment: A Review of Issues.* EPG Environmental Planning Issues series no. 11, London: International Institute for Environment and Development.

IIED. 1994. *Whose Eden? An Overview of Community Approaches to Wildlife Management.* London: International Institute for Environment and Development.

IIRR. 1998. *Participatory Methods in Community-based Coastal Resource Management.* Cavite, Philippines: International Institute for Rural Reconstruction.

Ingles, A. W. and A. S. Inglis. 1995. Data Gathering Activities of the Kibale and Semliki Conservation and Development Project, Uganda. Consultancy Report, IUCN, Switzerland.

Isaac, T. M. T. and K. N. Harilal. 1997. Planning for Empowerment: People's Campaign for Decentralised Planning in Kerala. *Economic and Political Weekly* 32, 1–2: 53–8.

Isaac, T. M. T. and P. K. M. Tharakan. 1995. Kerala: Towards a New Agenda. *Economic and Political Weekly* 30, 31 and 32: 1993–2004.

IUCN. 1997a. *Indigenous Peoples and Sustainability: Cases and Actions.* Utrecht: International Books.

IUCN. 1997b *Plan D'Aménagement du Parc National de Waza.* Maroua: Waza Logone National Park.

IUCN/UNEP/WWF. 1991. *Caring for the Earth, a Strategy for Sustainable Living.* Gland: United Nations.

Jackson, W. J. 1993. Action Research for Community Forestry. The Case of the Nepal Australia Community Forestry Project. Discussion Paper 3/93, Kathmandu: Nepal Australia Community Forestry Project.

Jackson, W. J. and D. Bond 1997. Monitoring and Evaluating Collaborative Management of Natural Resources in Eastern Africa. Workshop Report (7–11 April, 1997), Tanga Coastal Zone Conservation and Development Programme, Tanga, Tanzania.

Jacob, J. 1996. *Institutional Interventions in Water Management: A Micro Level Study of Organisation of Irrigation in Kerala.* Report of the Project *Towards Sustainable Development: An Actor-oriented Perspective.* Trivandrum: Centre for Development Studies.

Jacob, J. 1996a. *Group Action in Land Management: A Case Study of Group Farming of Paddy in Kerala.* Report of the Project *Towards Sustainable Development: An Actor-oriented Perspective.* Trivandrum: Centre for Development Studies.

Jentoft, S. 1989. Fisheries Co-management, *Marine Policy* 13: 137–54.

Jiggins, J. and H. de Zeeuw. 1992. Participatory Technology Development in Practice: Process and Methods, pp. 135–62 in *Farming for the Future: An Introduction to Low-External-Input and Sustainable Agriculture* (eds) C. Reijntjes, B. Haverkort, and A. Waters-Bayer. Leusden: ILEIA.

Jodha, N. S. 1990. Rural Common Property Resources: Contributions and Crisis. *Economic and Political Weekly* **25**, 26: 65–78.

Jonsson, S. and A. Rai (eds), 1994. *Forests, People and Protection: Case Studies of Voluntary Forest Protection by Communities in Orissa*. New Delhi: ISO/Swedforest and SIDA.

Jose, S. 1991. *Group Farming in Kerala: An Illustrative Study*. Unpublished M.Phil thesis, Trivandrum: Centre for Development Studies.

Joshi, A. 1998. Progressive Bureaucracy: An Oxymoron? The Case of Joint Forest Management in India. Brighton: Institute of Development Studies, University of Sussex.

Kant, S., N. M Singh and K. K Singh. 1991. *Community Based Forest Management Systems: Case Studies from Orissa*. New Delhi: ISO/ Swedforest Indian Institute of Forest Management and Swedish International Development Authority.

Karnataka Forest Department. 1994–96. *Project Process Support Team Documents, 1–39*. Swansea: Centre for Development Studies for Overseas Development Administration.

Keohane, R. O. and E. Ostrom (eds), 1995. *Local Commons and Global Interdependence: Heterogeneity and Cooperation in Two Domains*. Thousand Oaks, CA: Sage Publications.

Kidd, R. 1979. *Liberation or Domestication: Popular Theatre and Non-formal Education in Africa*. Educational Broadcasting International.

Kimber, R. 1981. Collective Action and the Fallacy of the Liberal Fallacy. *World Politics* **33**, 2: 178–96.

Knight, J. 1992. *Institutions and Social Conflict*. Cambridge: Cambridge University Press.

Korten, D. 1986. *Community Organisations and Rural Development: A Learning Process Approach*. New York: Ford Foundation.

Korten, D. C. (ed.) 1986. *Community Management: Asian Experience and Perspective*. West Hartford: Kumarian Press.

Krishnan, T. N. 1991. Wage Employment and Output in Interrelated Labour Market in Agrarian Economy: A Case of Kerala. *Economic and Political Weekly* **26**, 26: A86–A96.

Leach, M. and R. Mearns (eds), 1996. *The Lie of the Land: Challenging Received Wisdom on the African Environment*. Oxford: James Currey.

Leach, M., R. Mearns and I. Scoones (eds) 1997a. Community Based Sustainable Development: Consensus or Conflict? *IDS Bulletin* **28**, 4.

Leach, M., R. Mearns and I. Scoones. 1997b. *Environmental Entitlements. A Framework for Understanding the Institutional Dynamics of Environmental Change*. IDS Discussion Paper 359, Brighton: Institute of Development Studies, University of Sussex.

Lebeuf, A. 1969. *Les principautés Kotoko. Essai sur le caractère sacré de l'autorité*. Paris: CNRS.

Lele, U., K. Mitra and O. N. Kaul. 1994. *Environment, Development, and Poverty: A Report of the International Workshop on India's Forest Management and Ecological Revival*. CIFOR Occasional Paper No. 3, Bogor: Center for International Forestry Research.

Leurs, R. 1996. Current Challenges Facing Participatory Rural Appraisal. *Public Administration and Development* **16**, 1: 57–73.

Levang, P. 1992. *Pahmungan, Penengahan, Balai Kencana. Enquête Agro-économique dans la Région de Krui (Lampung)*. Research report, ORSTOM/BIOTROP.

Libecap, G. D. 1995. The Conditions for Successful Collective Action, pp. 161–89 in *Local Commons and Global Interdependence* (eds) R. O. Keohane and E. Ostrom. London: Sage Publishers.

Long, N. and A. Long (eds) 1992. *Battlefields of Knowledge: The Interlocking of Theory and Practice in Social Research and Development*. London and New York: Routledge.

Lovelace, G. W. 1984. Cultural Beliefs and the Management of Agro-ecosystems, pp. 194–205 in *An Introduction to Human Ecology Research on Agricultural Systems in Southeast Asia* (eds) A. T. Rambo and P. E. Sajise. Los Baños: University of the Philippines.

Lynch, O. and K. Talbott. 1996. *Balancing Acts: Community Based Forest Management and National Law in Asia and the Pacific*. Washington, D.C.: World Resources Institute.

Maesson, O. 1992. *Environmental Screening of NGO Development Projects*. Ottawa. CCIC.

Maheshwari, B. L. and A. H. Moosvi. n.d. *Institutional Development Study Andhra Pradesh, India*. Hyderabad: Ministry of Environment and Forests.

Maheshwari, J. K. 1990. Interaction of Tribals with the Forests, pp. 115–26. In *Tropical Forest Ecosystem Conservation and Development in South and South-East Asia: Proceedings MAB Regional Training Workshop*. Peechi: Kerala Forest Research Institute.

Maikhuri, R. K. and A. K. Gangwar. 1991. Fuelwood use by Different Tribal and Non-tribal Communities in North-east India. *Natural Resources Forum* **15**, 2: 162–5.

Maikhuri, R. K. 1991. Fuelwood Consumption Pattern of Different Tribal Communities Living in Arunachal Pradesh in North East India. *Bioresource Technology*. **35**: 291–6.

Mary, F. 1986. *Agroforets et Sociétés: Etude Comparée de Trois Systèmes Agroforestiers Indonésiens*. Unpublished PhD thesis, ENSA-Montpellier, France.

Mary, F. and G. Michon. 1987. When Agroforests drive back Natural Forests: A Socio-economic Analysis of a Rice/agroforest System in South Sumatra. *Agroforestry Systems* **5**: 27–55.

McCay, B. J. and J. M. Acheson (eds), 1990. *The Question of the Commons: The Culture and Ecology of Communal Resources*. Tucson: University of Arizona Press.

McCracken, J. A., J. N. Pretty and G. R. Conway. 1988. *An Introduction to Rapid Rural Appraisal for Agricultural Development*. London: International Institute for Environment and Development.

McNeely, J. A. 1996. *Conservation and Future: Trends and Options Toward the Year 2025*. Discussion paper. Gland: IUCN, The World Conservation Union.

Michon, G. 1985. *De l'homme de la Forêt au Paysan de l'Arbre: Agroforesteries Indonésiennes*. Unpublished PhD Thesis, U.S.T.L., Montpellier, France.

Michon, G., F. Mary and J. M. Bompard. 1986. Multistoried Agroforestry Garden System in West Sumatra, Indonesia. *Agroforestry Systems* **4**: 315–38.

Michon, G. and J. M. Bompard. 1987. Agroforesteries Indonésiennes: Contributions Paysannes à la Conservation des Forêts Naturelles et de leurs Ressources. *Rev. Ecol. (Terre Vie)* **42**: 3–37.

Michon, G. and H. de Foresta. 1990. Complex Agroforestry Systems and Conservation of Biological Diversity: Agroforestry in Indonesia, A Link between Two Worlds. *The Malayan Nature Journal* **45**: 457–73.

Michon, G. and H. de Foresta. 1995. The Indonesian Agro-forest Model, pp. 90–106 in *Conserving Biodiversity outside Protected Areas: The Role of Traditional Ecosystems* (eds) P. Halladay and D. A. Gilmour. IUCN: Gland, Switzerland and Cambridge, UK: IUCN, The World Conservation Union.

Michon, G., H. de Foresta and P. Levang. 1995. Stratégies Agroforestières Paysannes et Développement Durable: Les Agroforêts à *damar* de Sumatra. *Nature-Sciences-Sociétés* **3**, 3: 207–21.

Midgley, J. 1986. Community Participation: History, Concepts and Controversies. In *Community Participation, Social Development and the State* (eds) J. Midgley, A. Hall, M. Hardiman and D. Narine. London: Methuen.

Ministry of Environment and Forests. 1988. *National Forest Policy Resolution 1988 (Initiatives in Conservation of Forests)*. New Delhi: Ministry of Environment and Forests.

Ministry of Environment and Forests. 1990. *Circular on Involvement of Village Communities and Voluntary Agencies for Regeneration of Degraded Forest Lands.* Ministry of Environment and Forests 6–21/89–F.P.

Momberg, F. 1993. *Indigenous Knowledge Systems: Potentials for Social Forestry Development: Resource Management of Land-Dayaks in West Kalimantan.* Unpublished Masters Thesis, Berlin: Technische Universität Berlin.

Mosse, D. 1994. Authority, Gender and Knowledge: Theoretical Reflections on the Practice of Participatory Rural Appraisal. *Development and Change* **25**, 3: 497–525.

Mosse, D. 1995. Authority, Gender and Knowledge: Theoretical Reflections on Participatory Rural Appraisal. *Economic and Political Weekly* **30**, 11: 569–78.

Mulekom, L. van. 1998. *Capacity and Opportunity Building in a Small-scale Fisherfolk Community for Community-based Coastal Resource Management: a Community Development Approach to CB-CRM* (draft) Philippines: SNV.

Mulekom, L. van and E. C. Tria. 1997. Community-based Coastal Resource Management in Orion (Bataan, Philippines): Building Property Rights in a Fishing Community. *NAGA* April–June 1997: 51–5.

Narayan, D. 1995. *The Contribution of People's Participation: Evidence from 121 Rural Water Supply Projects.* Washington D.C.: The World Bank.

Narayana, D. 1992. *Interaction of Price and Technology in the Presence of Structural Specificities: An Analysis of Crop Production in Kerala.* Unpublished PhD thesis, Calcutta: Indian Statistical Institute.

NCAER. 1962. *Techno-Economic Survey: Kerala*. New Delhi: National Council for Applied Economic Research.

Neefjes, K. 1992. Approaches to Environment and Development in the Strategic Plans. Oxfam internal mimeo 10.92, Oxford: Oxfam.

Neefjes, K. 1998a. Oxfam's Impact on Livelihoods in Lung Vai – A Study of Change in a Commune in Lao Cai Province, Vietnam. Unpublished mimeo, Hanoi/Oxford: Oxfam.

Neefjes, K. 1998b. Ecological Needs of Communities During and After Dryland Crises in *Plants for Food and Medicine* (eds) H. D. V. Prendergast, N. L. Etkin, D. R. Harris and P. J. Houghton. Kew: Royal Botanic Gardens.

Neefjes, K. 1998c. Food Security in Southern Niassa – A Mid-term Review of the Impact of Oxfam's Programme in Nipepe, Metarica and Maua Districts. Unpublished mimeo, Cuamba/Nipepe/Oxford: Oxfam.

Neefjes, K. 1999a. *Participatory Review in Chronic Instability – The Experience of the 'Ikafe' Refugee Settlement Programme, Uganda*. London: Overseas Development Institute.

Neefjes, K. 1999b. Ecological Degradation: A Cause for Conflict, a Concern for Survival in *Fairness and Futurity* (ed.) A. Dobson. Oxford: Oxford University Press.

Neefjes, K. and B. Nakacwa. 1996. To PEA or Not to PEA? Assessment of Learning Needs on Environment and Development of Some Ugandan NGOs. Unpublished mimeo, Oxfam/Novib/SNV and Environmental Alert.

Neefjes, K., P. Mafongoya and M. Mwangi, with E. Ngunjiri and E. Mugure. 1997. Conservation Farming, Food Security and Social Justice – A Sectoral Review of Agricultural Work with Small NGOs and National Networking by Oxfam-Kenya. Unpublished mimeo, Nairobi/Oxford: Oxfam.

Neefjes. K. and W. Woldegiorgis. 1999. Livelihoods and Staff Capacity: An Assessment of Impacts of Training. Unpublished mimeo. Oxford: Oxfam GB, Gender and Learning Team.

Neetha, N. forthcoming. *Irrigation Institutions in Kerala: A Study of the Chalakudy Project*. Doctoral Thesis, Trivandrum: Centre for Development Studies.

Nelson, D. and E. Silberberg. 1987. Ideology and Legislator Shirking. *Economic Inquiry* 25: 15–25.

Nelson, N. and S. Wright (eds) 1995. *Power and Participatory Development: Theory and Practice*. London: Intermediate Technology Publications.

North, D. C. 1990. *Institutions, Institutional Change and Economic Performance*. Cambridge: Cambridge University Press.

Novaczek, I. and I. Harkes. 1998. *An Institutional Analysis of Sasi Laut in Maluku, Indonesia*. Working Paper No. 39, Manila: ICLARM.

Nurse, M. C. 1996. Interim Report of the Consultancy to Assess the Coastal Forest Resources in the Tanga Region. Tanga Coastal Zone Conservation and Development Programme, Tanga, Tanzania.

Nurse, M. C. 1997. Coastal Forest Assessment Consultancy, Final Report. Tanga Coastal Zone Conservation and Development Programme, Tanga, Tanzania.

Nurse, M. C. 1998. Coastal Forest Assessment Consultancy, Second Report, Phase II. Tanga Coastal Zone Conservation and Development Programme, Tanga, Tanzania.

Nurse, M. C. in press. Forest Management Arrangements. In *Non-Timber Forest Product Utilisation of the Ritigala Strict Natural Reserve, Sri Lanka: Balancing Livelihood Strategies with Conservation Management* (eds) A. Wickramasinghe, M. Ruiz-Perez and J. Blockhus. Issues in Forest Conservation Series, Gland: IUCN, The World Conservation Union.

Nurse, M. C., C. R. Mc Kay, J. B. Young, and C. A. Asanga. 1994. Biodiversity Conservation through Community Forestry, in the Montane Forests of Cameroon. Presented at the *BirdLife International XXI World Conference: Global Partnership for Bird Conservation* 12–18 August 1994. Rosenheim, Germany.

Oakley, P. and D. Marsden. 1984. *Approaches to Participation in Rural Development*. ACC Task Force on Rural Development, Geneva: International Labour Office.

ODA. 1996. *ODA's Review of Participatory Forest Management: Synthesis of Findings*. London: Overseas Development Administration.

ODI. 1998. *A Manual on Alternative Conflict Management for Community-based Natural Resource Projects in the South Pacific*. London: Overseas Development Institute.

Olivier de Sardan, J.-P. 1988. Peasant Logics and Development Project Logics. *Sociologica Ruralis* **28**, 2/3: 216–26.

Olson, M. 1965. *The Logic of Collective Action*. Cambridge, MA: Harvard University Press.

Oommen, M. A. 1963. The Economics of Cropping Pattern: A Case Study. *The Indian Journal of Agricultural Economics* **18**, 1: 120–8.

Ostrom, E. 1990. *Governing the Commons: The Evolution of Institutions for Collective Action*. Cambridge: Cambridge University Press.

Ostrom, E. 1992. Community and the Endogenous Solution of Commons Problems. *Journal of Theoretical Politics* **4**, 3: 343–51.

Ostrom, E. 1994. Constituting Social Capital and Collective Action. *Journal of Theoretical Politics* **6**: 527–62.

Ostrom, E. 1996. Crossing the Great Divide: Coproduction, Synergy, and Development. *World Development* **24**, 6: 1073–87.

Ostrom, E. 1999. *Self-Governance and Forest Resources*. CIFOR Occasional Paper No 20. Bogor: Center for International Forestry Research.

Ostrom, V. 1996. Syllabus: A Course of Study in Institutional Analysis and Development: Political Order and Development: Macro (Political Science Y673). First Semester, Fall Term 1996. Bloomington: Workshop in Political Theory and Policy Analysis, Indiana University. [http://www.indiana.edu/~workshop/fallsem.html].

Oxfam. 1992. *Strategic Plan FY 1992/93–1994/95*. Overseas Division. Oxford: Oxfam.

Oxfam. 1995. *Guidelines for Strategic Planning, Action Planning and Reporting*. Overseas Division. Oxford: Oxfam.

Oxfam GB. 1999 *Oxfam (GB)'s Work With Agricultural Communities in Kenya: Promoting Food Security and Social Justice Through Conservation Framing. A Review of Project Impact*, Report for the Department for International Development. Oxford: Oxfam.

Pacey, A. 1982. *The Culture of Technology*. Oxford: Blackwell.

Palmer, R. 1997. Contested Lands in Southern and Eastern Africa: A Literature Survey. Oxfam Working Paper, Oxford: Oxfam.

Pant, N. 1999. Impact of Irrigation Management Transfer in Maharashtra: An Assessment. *Economic and Political Weekly* **34**, 13: A17–A26.

Pasicolan, P. 1996. *Tree Growing in Different Grounds: An Analysis of Local Participation in Contract Reforestation in the Philippines*. Unpublished PhD dissertation, Leiden: Leiden University.

Pathak, A. 1994. *Contested Domains: The State, Peasants and Forests in Contemporary India*. New Delhi, Sage Publications.

Patnaik, U. 1996. Export-Oriented Agriculture and Food Security in Developing Countries and India. *Economic and Political Weekly* **31**: 2429–50.

Paul, S. 1987. *Community Participation in Development Projects: The World Bank Experience*. World Bank Discussion Paper No. 6. Washington, D.C.: The World Bank.

Peet, R. and M. Watts (eds) 1996. *Liberation Ecologies: Environment, Development, Social Movements*. London and New York: Routledge.

Peluso, N. L. 1992. *Rich Forests, Poor People: Resource Control and Resistance in Java*. Berkeley: University of California Press.

Persoon, G. and P. Sajise. 1997. Co-management of Natural Resources in Asia: A Comparative Perspective. Paper presented at IIAS/NIAS Seminar, Philippines.

Pido, M. D., R. S. Pomeroy, M. B. Carlos and L. R. Garces. 1996. *A Handbook for Rapid Appraisal of Fisheries Management Systems*. Manila: ICLARM.

Pike, K. 1954. *Language in Relation to a Unified Theory of the Structure of Human Behavior (Vol. 1)*. Glendale: Summer Institute of Linguistics.

Pimbert, M. and J. Pretty. 1997. Diversity and Sustainability in Community-Based Conservation. Paper for the UNESCO-IIPA regional workshop on Community-based Conservation, 9–12 February, India.

Pinkerton, E. (ed.) 1989. *Co-operative Management of Local Fisheries: New Directions for Improved Management and Community Development*. Vancouver: University of British Columbia Press.

Poffenberger, M. 1990a. Facilitating Change in Forest Bureaucracies. In *Keepers of the Forest* (ed.) M. Poffenberger. West Hartford, CT: Kumarian Press.

Poffenberger, M. 1990b. Joint Management for Public Forests: Experiences from South Asia, pp. 158–82 in *Research Policy for Community Forestry, Asia Pacific Region*. Bangkok: Regional Community Forestry Training Center.

Poffenberger, M. 1990c. *Keepers of the Forest: Land Management Alternatives in Southeast Asia*. West Hartford: Kumarian Press.

Poffenberger, M. 1995. Public Lands Reform: India's Experience with Joint Forest Management. Unpublished Mimeo, Berkeley: Asia Forest Network, Center for Southeast Asia Studies.

Poffenberger, M. and B. Mc Gean (eds) 1996. *Village Voices, Forest Choices: Joint Forest Management in India*. Delhi: Oxford University Press.

Poffenberger, M. *et al.* 1996. *Grassroots Forest Protection: Eastern India Experiences*. Asia Forest Network Research Report No. 7, San Francisco: Asia Forest Network.

Pollnac, R. B. 1989. *Monitoring and Evaluating the Impacts of Small-scale Fishery Projects*. Rhode Island: ICMRD.

Pollnac, R. B. and J. J. Poggie 1988. The Structure of Job Satisfaction among New England Fishermen and its Application to Fisheries Management Policy. *American Anthropologist* 90: 888–901.

Pomeroy, R. S. (ed.) 1994. *Community Management and Common Property of Fisheries in Asia and the Pacific: Concepts, Methods and Experiences*. Manila: ICLARM.

Pomeroy, R. S. 1998. A Process for Community-based Fisheries Co-management. *NAGA*, January–March 1998.

Pomeroy, R. S. and F. Berkes. 1997. Two to Tango: The Role of Government in Fisheries Co-management. *Marine Policy* 21, 5: 465–80.

Pomeroy, R. S. and M. B. Carlos. 1996. A Review and Evaluation of Community-based Coastal Resources Management Projects in the Philippines, 1984–1994. Research Report No. 6, Manila: ICLARM.

Pomeroy, R. S. and M. B. Carlos. 1997. Community-based Coastal Resource Management in the Philippines: A Review and Evaluation of Programs and Projects, 1984–1994. *Marine Policy* 21, 2: 445–64.

Pomeroy, R. S., R. B. Pollnac, C. D. Predo and B. M. Katon. 1996. Impact Evaluation of Community-based Coastal Resource Management Projects in the Philippines. Research Report No. 3, Manila: ICLARM.

Potter, D. 1998. NGOs and Forest Management in Karnataka. In *The Social Construction of Indian Forests* (ed.) R. Jeffery. Edinburgh and Delhi: Centre for South Asian Studies and Manohar Publishers.

Pretty, J. 1994. Alternative Systems of Inquiry for a Sustainable Agriculture. *IDS Bulletin* 25, 2: 37–48.

Putnam, R. (with R. Leonardi and R. Y. Nanetti). 1993. *Making Democracy Work: Civic Traditions in Modern Italy*. Princeton: Princeton University Press.

Putnam, R. 1995. 'Bowling Along: America's Declining Social Capital, *Journal of Democracy*. 6, 65–78.

Pye-Smith, C. and G. Borrini-Feyerabend, with R. Sandbrook. 1994. *The Wealth of Communities: Stories of Success in Local Environmental Management*. London: Earthscan.

Rajan, R. 1998. Imperial Environmentalism or Environmental Imperialism? European Forestry, Colonial Foresters and the Agendas of Forest Management in British India 1800–1900. In *Nature and the Orient* (eds) R. Grove, V. Damodaran and S. Sangwan. Delhi: Oxford University Press.

Ramamani, V. S. 1988. *Tribal Economy: Problems and Prospects*. Allahabad: Chugh Publications.

Rao, R. and B. P. Singh. 1996. Non-wood Forest Products Contribution in Tribal Economy. *Indian Forester* 122: 337–41.

Resolve, 1994. *The Role of Alternative Conflict Management in Community Forestry*. Community Forestry Working Paper No. 1, Rome: Food and Agriculture Organisation of the United Nations.

Roche, C. 1999. *Impact Assessment for Development Agencies: Learning to Value Change*. Oxford: Oxfam.

Roy Burman, J. J. 1990. A Need for Reappraisal of Minor Forest Produce Policies. *The Indian Journal of Social Work* 51: 649–58.

Runge C. F. 1984. Institutions and the Free-Rider: The Assurance Problem in Collective Action. *The Journal of Politics* 46: 154–81.

Runge, C. F. 1992. Common Property and Collective Action in Economic Development, pp. 17–40 in *Making the Commons Work: Theory, Practice and Policy* (ed) D. W. Bromley. California: ICS Press.

Saigal, S., C. Agarwal and J. Campbell. 1996. *Sustaining Forest Management: The Role of Non-Timber Forest Products*. New Delhi: Society for the Promotion of Wastelands Development.

Sajise, P. 1995. Community-based Resource Management in the Philippines: Perspectives and Experiences. Paper presented at the Fisheries Co-management Workshop at the North Sea Centre, Hirtshals, Denmark, 29–31 May.

Salafsky, N. 1993. *The Forest Garden Project: An Ecological and Economic Study of a Locally Developed Land-Use System in West Kalimantan, Indonesia*. Unpublished PhD Thesis, Durham, NC: Duke University.

Santhakumar, V. and R. Rajagopalan. 1995. The Green Revolution in Kerala: A Discourse on Technology and Nature. *South Asia Bulletin* 15, 2: 109–19.

Santhakumar, V., R. Rajagopalan and S. Ambirajan. 1995. Planning Kerala's Irrigation Projects: Technological Prejudice and Politics of Hope. *Economic and Political Weekly* 30, 12: A30–A38.

Sardjono, M. A. 1988. Lembo: A Traditional Land-use System in East Kalimantan, pp. 253–66 in *Agroforestry Untuk Pengembangan Daerah Pedesaan di Kalimantan*

Timur, (eds) A. M. Lahjie and B. Seibert. Fakultas Kehutanan Universitas Mulawarman and GTZ.

Sarin, M. 1996. *Actions of the Voiceless: The Challenge of Addressing Subterranean Conflicts Related to Marginalised Groups and Women in Community Forestry.* Paper presented in the FAO E-Conference on Conflict Management, January–April.

Sarin, M. 1998. *Who is Gaining? Who is Losing? Gender and Equity Concerns in Joint Forest Management.* New Delhi: Society for the Promotion of Wastelands Development.

Saxena, N. C. 1997. *The Saga of Participatory Forest Management in India.* Bogor: Center for International Forestry Research.

Saxena, N. C. n.d. *The Indian Forest Service.* Mussoorie: Centre for Sustainable Development, Lal Bahadur Shastri National Academy of Administration.

Scheinman, D and A. Mabrook 1996. The Traditional Management of Coastal Resources. Consultancy Report, Tanga Coastal Zone Conservation and Development Programme, Tanzania.

Scholte, P., S. Kari and M. Moritz. 1996. *The Involvement of Nomadic and Transhumant Pastoralists in the Rehabilitation and Management of the Logone Flood Plain, North Cameroon.* Issues Paper no. 66, London, International Institute for Environment and Development.

Scoones, I. 1998. *Sustainable Rural Livelihoods: A Framework for Analysis.* IDS working paper no. 72, Brighton: Institute of Development Studies, University of Sussex.

Scott, P. 1994. Resource Use Assessment Report. Mount Elgon National Park, Uganda. Consultancy Report, Gland: IUCN, The World Conservation Union.

Scott, P. in preparation. *Collaborative Management: A Case Study from the Ruwenzori Mountains National Park, Uganda.* Uganda: World Wide Fund for Nature.

Shepherd, G. 1996. Local Systems of Resource Control and Links to Policy, pp. 43–50. In *Making Forest Policy Work* (ed.) K. Harris. Conference Proceedings of the Oxford Summer Course Programme, Oxford: Oxford Forestry Institute.

Sibuea, T. T. H. and D. Herdimansyah. 1993. *The Variety of Mammal Species in the Agroforest Areas of Krui (Lampung), Muara Bungo (Jambi), and Maninjau (West Sumatra).* Research report, Bandung: ORSTOM/HIMBIO (UNPAD).

Siebert, S. F. 1989. The Dilemma of Dwindling Resource: Rattan in Kerinci, Sumatra. *Principes* **32**, 2: 79–97.

Singh, S. and A. Khare. 1993. Peoples' Participation in Forest Management. *Wastelands News* III: 34–8.

Socpa, A. 1992. Origines Socio-historique des Heurts Interethniques Arab-Shoa – Kotoko. In *Les affrontements Interethniques entre Arabes Choa et Kotoko dans le Logone et Chari* (eds) Tsala and Onambele. Yaoundé: Rapport Tribus sans Frontières no. I.

Spierenburg, M. 1997. *Development a Matter of Time: Social Memory and the Struggle for Control over Land in Dande, Northern Zimbabwe.* Paper presented at the African PDO seminar, Leiden: Leiden University.

SPWD. 1993. *Joint Forest Management Regulations Update.* New Delhi: Society for the Promotion of Wasteland Development.

Stubbs, J. and N. MacDonald. 1993. South-South Environment Linking Project – A Report of the Project up to January 1993. Unpublished mimeo, Oxford: Oxfam.

Sundar, N. *et al.* 1996. Defending the Dalki Forest – 'Joint' Forest Management in Lapanga. *Economic and Political Weekly* **31**: 3021–25.

Sundawati, L. 1993. *The Dayak Garden Systems in Sanggau District, West Kalimantan. An Agroforestry Model.* Unpublished MSc. thesis, Göttingen: Faculty of Forestry, Georg-August University.

Taylor, M. 1982. *Community, Anarchy and Liberty.* Cambridge: Cambridge University Press.

TCZC and DP. n.d. Tanga Coastal Zone Conservation and Development Programme – Fact Sheet.

Tewari, D. N. 1989. *Dependence of Tribals on Forests.* Ahmedabad: Gujarat Vidyapith.

Thin, N., N. Peter and P. Gorada. 1998. *Muddles about the Middle: NGOs as Intermediaries in JFM.* (Edinburgh Papers in South Asian Studies 9). Edinburgh: Centre for South Asian Studies.

Thiollay, J. M. 1995. The Role of Traditional Agroforests in the Conservation of Rain Forest Bird Diversity in Sumatra. *Conservation Biology* **9**, 2: 335–53.

Thomas, A., J. Chataway and M. Wuyts (eds) 1998. *Finding Out Fast: Investigative Skills for Policy and Development.* London/Thousand Oaks/New Delhi: Sage Publications (in association with The Open University).

Thomas, D. H. L. and W. M. Adams. 1997. Space, Time and Sustainability in the Hadejia-Jama'are Wetlands and the Komodugu Yobe Basin, Nigeria. *Transactions of the Institute of British Geographers* NS **22**: 430–49.

Thompson, J. 1995. Participatory Approaches in Government Bureaucracies: Facilitating the Process of Institutional Change. *World Development* **23**: 1521–54.

Thompson, J. and K. S. Freudenberger. 1997. *Crafting Institutional Arrangements for Community Forestry.* FAO Community Forestry Field Manual 7. Rome: Food and Agriculture Organisation of the United Nations.

Torquebiau, E. 1984. Man-made Dipterocarp Forest in Sumatra. *Agroforestry Systems* **2**, 2: 103–28.

UNDP. 1993. *Tools for Community Participation. A Manual for Training Trainers in Participatory Techniques.* Washington: PROWWESS/UNDP-World Bank Water and Sanitation Program.

UNDP. 1997. *UNDP Guidebook on Participation.* New York: United Nations Development Programme.

Uphoff, N. T. 1982. *Rural Development and Local Organisations in Asia.* New Delhi: Macmillan India Ltd.

Uphoff, N. T. 1986. *Local Institutional Development: An Analytical Sourcebook with Cases.* West Hartford: Kumarian Press.

Uphoff, N. T. 1988. Assisted Self-Reliance: Working with, Rather than for, the Poor. In *Strengthening the Poor: What Have We Learned?* (ed.) J. P. Lewis. New Brunswick: Transaction Books.

Uphoff, N. T. 1992. *Local Institutions and Participation for Sustainable Development.* Gatekeeper series No. 31, London: International Institute for Environment and Development.

Vaidyanathan, A. 1996. Agricultural Development: Imperatives of Institutional Reforms. *Economic and Political Weekly* **31**, 35–37: 2451–8.

Vira, B. 1997. *Deconstructing Participatory Forest Mangement: Toward A Tenable Typology.* OCEES Research Paper No. 14, Oxford: Oxford Centre for the Environment, Ethics and Society.

Vira, B. 1999. Implementing JFM in the Field: Towards an Understanding of the Community-Bureaucracy Interface. In *A New Moral Economy for India's Forests?* (eds) R. Jeffery and N. Sundar. New Delhi: Sage.

Vira, B., O. Dubois, S. E. Daniels and G. B. Walker. 1998. Institutional Pluralism in Forestry: Considerations of Analytical and Operational Tools. *Unaslyva* **49**, 194: 35–42.

Wallman, S. (ed.) 1992. *Contemporary Futures. Perspectives from Social Anthropology.* London: Routledge.

Warner, M. and P. Jones. 1998. *Assessing the Need to Manage Conflict in Community-Based Natural Resource Projects.* Natural Resource Perspectives No. 35, London: Overseas Development Institute.

Watkins, K. 1995. *The Oxfam Poverty Report.* Oxford: Oxfam.

Weimer, D. L. and A. R. Vining. 1992. *Policy Analysis: Concepts and Practice.*Englewood Cliffs, NJ: Prentice Hall.

Weinstock, J. A. 1983. Rattan: Ecological Balance in a Borneo Rainforest Swidden. *Economic Botany* **37**, 1: 58–68.

Western, D. 1994. Vision of the Future. The New Focus of Conservation, pp. 548–56 in *Natural Connections: Perspectives in Community-based Management* (eds) D. Western and R. Wright. Washington, D.C.: Island Press.

Western, D. and R. Wright (eds) 1994. *Natural Connections: Perspectives in Community-based Management.* Washington, D.C.: Island Press.

Wiersum, F. 1996. Indigenous Exploitation and Management of Tropical Forest Resources: An Evolutionary Continuum in Forest–People Interaction. *Agriculture Ecosystems and Environment* **63**: 1–16.

Wilterdink, N. and B. van Heerikhuizen. 1985. *Samenlevingen: een Verkenning van het Terrein van de Sociologie.* Groningen: Wolters-Noordhoff.

Wily, E. 1995. The Emergence of Joint Forest Management in Tanzania – Villager and Government: The Case of Mgori Forest, Singida Region. Unpublished report.

Wily, E. 1996. Good News from Tanzania! The First Village Forest Reserves in the Making. The Story of Duru-Haitenba Forest. *Forest, Trees and People Newsletter* **29**: 29–37.

World Bank. 1997. India to Protect Kerala's Forest Resources with help from Local Communities. Press release No. 98/1692/SAS, Washington, D.C.: The World Bank.

World Bank. 1998. Kerala Forestry Project: Project Appraisal Document. Rural Development Sector Unit, South Asia Region, Washington, D.C.: The World Bank.

World Commission on Environment and Development. 1987. *Our Common Future.* Oxford: Oxford University Press.

WWF. 1996. *Indigenous Peoples and Conservation: WWF Statements and Principles.* Gland: World Wide Fund for Nature.

Yadama, G. N., B. R. Pragada and R. R. Pragada. 1997. *Forest Dependent Survival Strategies of Tribal Women: Implications for Joint Forest Management in Andhra Pradesh, India.* Bangkok: Food and Agriculture Organization of the United Nations.

Index